STO

CHANGES IN THE 1981 NATIONAL ELECTRICAL CODE®

CHANGES IN THE 1981 NATIONAL ELECTRICAL CODE®

George W. Flach, P.E.

PRENTICE-HALL, INC., *Englewood Cliffs, New Jersey 07632*

ALLEN COUNTY PUBLIC LIBRARY
FORT WAYNE, INDIANA

Library of Congress Cataloging in Publication Data

Flach, George W
  Changes in the 1981 National electrical code®.

  1. Electric engineering—Insurance requirements.
  2. National Fire Protection Association. National
electrical code, 1981. I. Title.
TK260.F57   621.319'24'0218   80-27386
ISBN 0-13-127860-6
ISBN 0-13-127852-5 (pbk.)

®"NATIONAL ELECTRICAL CODE, NEC are trademarks of
National Fire Protection Association, Inc., Boston,
Massachusetts, for a triannual electrical publication.
The terms NATIONAL ELECTRICAL CODE, NEC, as used herein
means the triannual publication constituting the
NATIONAL ELECTRICAL CODE and is used with permission
of National Fire Protection Association, Inc."

Reproduced by permission from the 1981 National Electrical Code®
Copyright © 1980, National Fire Protection Association, Boston, MA.

Editorial/production supervision by Ellen De Filippis
Manufacturing buyer: Gordon Osbourne

© 1981 by Prentice-Hall, Inc., Englewood Cliffs, N.J. 07632

All rights reserved. No part of this book
may be reproduced in any form or
by any means without permission in writing
from the publisher.

Printed in the United States of America

10 9 8 7 6 5 4 3 2 1

PRENTICE-HALL International, Inc., *London*
PRENTICE-HALL of Australia Pty. Limited, *Sydney*
PRENTICE-HALL of Canada, Ltd., *Toronto*
PRENTICE-HALL of India Private Limited, *New Delhi*
PRENTICE-HALL of Japan, Inc., *Tokyo*
PRENTICE-HALL of Southeast Asia Pte. Ltd., *Singapore*
WHITEHALL BOOKS LIMITED, Wellington, *New Zealand*

Contents

Preface

xi

90 Introduction — *1*

ONE

General

3

100	Definitions	*5*
110	Requirements for Electrical Installations	*7*

TWO

Wiring Design and Protection

9

200	Use and Identification of Grounded Conductors	*11*
210	Branch Circuits	*12*
215	Feeders	*14*

220	Branch-Circuit and Feeder Calculations	*15*
225	Outside Branch Circuits and Feeders	*17*
230	Services	*19*
240	Overcurrent Protection	*24*
250	Grounding	*26*
280	Surge Arresters	*34*

THREE

Wiring Methods and Materials

37

300	Wiring Methods	*39*
305	Temporary Wiring	*41*
310	Conductors for General Wiring	*42*
318	Cable Trays	*46*
320	Open Wiring on Insulators	*47*
321	Messenger Supported Wiring	*47*
324	Concealed Knob-and-Tube Wiring	*48*
326	Medium Voltage Cable	*49*
328	Flat Conductor Cable Type FCC	*49*
330	Mineral-Insulated, Metal-Sheathed Cable	*52*
333	Armored Cable	*52*
334	Metal-Clad Cable	*52*
336	Nonmetallic-Sheathed Cable	*52*
337	Shielded Nonmetallic-Sheathed Cable	*53*
338	Service-Entrance Cable	*53*
339	Underground Feeder and Branch-Circuit Cable	*53*
340	Power and Control Tray Cable	*54*
342	Nonmetallic Extensions	*54*
344	Underplaster Extensions	*54*
345	Intermediate Metal Conduit	*54*
346	Rigid Metal Conduit	*55*
347	Rigid Nonmetallic Conduit	*55*
348	Electrical Metallic Tubing	*55*
349	Flexible Metallic Tubing	*56*
350	Flexible Metal Conduit	*56*
351	Liquidtight Flexible Conduit	*56*
352	Surface Raceways	*57*
353	Multioutlet Assembly	*57*
354	Underfloor Raceways	*57*
356	Cellular Metal Floor Raceways	*58*
358	Cellular Concrete Floor Raceways	*58*

Contents · vii

362	Wireways	58
363	Flat Cable Assemblies	59
364	Busways	59
365	Cablebus	59
366	Electrical Floor Assemblies	59
370	Outlet, Switch and Junction Boxes, and Fittings	60
373	Cabinets and Cutout Boxes	61
374	Auxiliary Gutters	63
380	Switches	63
384	Switchboards and Panelboards	64

FOUR

Equipment For General Use

67

400	Flexible Cords and Cables	69
402	Fixture Wires	70
410	Lighting Fixtures, Lampholders, Lamps, Receptacles and Rosettes	71
422	Appliances	74
424	Fixed Electric Space Heating Equipment	75
426	Fixed Outdoor Electric De-Icing and Snow-Melting Equipment	77
427	Fixed Electric Heating Equipment for Pipelines and Vessels	80
430	Motors, Motor Circuits, and Controllers	84
440	Air-Conditioning and Refrigerating Equipment	94
445	Generators	94
450	Transformers and Transformer Vaults	95
460	Capacitors	97
470	Resistors and Reactors	98
480	Storage Batteries	98

FIVE

Special Occupancies

99

500	Hazardous (Classified) Locations	101
501	Class I Locations	103

502	Class II Locations	*104*
503	Class III Locations	*106*
510	Hazardous (Classified) Locations—Specific	*106*
511	Commercial Garages, Repair and Storage	*106*
513	Aircraft Hangars	*106*
514	Gasoline Dispensing and Service Stations	*106*
515	Bulk Storage Plants	*109*
516	Finishing Processes	*110*
517	Health Care Facilities	*110*
518	Places of Assembly	*120*
520	Theaters and Similar Locations	*120*
530	Motion Picture and Television Studios and Similar Locations	*121*
540	Motion Picture Projectors	*121*
545	Manufactured Building	*121*
547	Agricultural Buildings	*121*
550	Mobile Homes and Mobile Home Parks	*121*
551	Recreational Vehicles and Recreational Vehicle Parks	*122*
555	Marinas and Boatyards	*124*

SIX

Special Equipment

125

600	Electric Signs and Outline Lighting	*127*
604	Manufactured Wiring Systems	*128*
610	Cranes and Hoists	*130*
620	Elevators, Dumbwaiters, Escalators, and Moving Walks	*130*
630	Electric Welders	*131*
640	Sound-Recording and Similar Equipment	*131*
645	Data Processing Systems	*131*
650	Organs	*132*
660	X-Ray Equipment	*132*
665	Induction and Dielectric Heating Equipment	*132*
668	Electrolytic Cells	*132*
669	Electroplating	*132*
670	Metalworking Machine Tools	*133*
675	Electrically Driven or Controlled Irrigation Machines	*134*
680	Swimming Pools, Fountains, and Similar Installations	*134*
685	Integrated Electrical Systems	*138*

SEVEN

Special Conditions

139

700	Emergency Systems	*141*
701	Legally Required Standby Systems	*144*
702	Optional Standby Systems	*145*
710	Over 600 Volts, Nominal—General	*146*
720	Circuits and Equipment Operating at Less than 50 Volts	*147*
725	Class 1, Class 2, and Class 3 Remote-Control, Signaling, and Power-Limited Circuits	*147*
760	Fire Protective Signaling Systems	*149*

EIGHT

Communications Systems

151

800	Communication Circuits	*153*
810	Radio and Television Equipment	*154*
820	Community Antenna Television and Radio Distribution Systems	*154*

NINE

Tables and Examples

157

Tables	*159*
Examples	*160*

Preface

This book explains the most important changes in the 1981 Edition of the National Electrical Code®. Where appropriate, reasons for the revisions are explained, and the impact they will have on the manufacture, installation, and inspection of electrical materials and equipment is pointed out.

Since the exact wording of each change is not mentioned in this text, the revision should first be read as it appears in the NEC, then the explanation should be reviewed. In this way the reader will have a clearer understanding of what the revision is all about.

All revisions, many editorial or the result of Technical Committee activity, are not mentioned because they have no significant effect on the electrical industry. Examples of these kinds of revisions are the soft conversion to the metric system and editorial changes to improve sentence structure.

This manuscript was prepared from the *"Preprint" of the Proposed Amendments for the 1981 National Electrical Code*® (National Electrical Code Technical Committee Report) and the *National Electrical Code Technical Committee Documentation,* together with actions taken at the NFPA annual meeting in Boston on May 19–22, 1980, so that its publication date would coincide as closely as possible with publication of the 1981 Edition of the National Electrical Code®.

All explanations and interpretations are the personal opinion of the author and have no official or legal status. As always, the final decision on interpretation of any requirements is the sole responsibility of the authority having jurisdiction. Requests for Official Interpretations of any of the rules should be addressed to the National Fire Protection Association.

George W. Flach

CHANGES IN THE 1981 NATIONAL ELECTRICAL CODE®

CHANGES IN THE 1981 NATIONAL ELECTRICAL CODE®

Preface

This book explains the most important changes in the 1981 Edition of the National Electrical Code®. Where appropriate, reasons for the revisions are explained, and the impact they will have on the manufacture, installation, and inspection of electrical materials and equipment is pointed out.

Since the exact wording of each change is not mentioned in this text, the revision should first be read as it appears in the NEC, then the explanation should be reviewed. In this way the reader will have a clearer understanding of what the revision is all about.

All revisions, many editorial or the result of Technical Committee activity, are not mentioned because they have no significant effect on the electrical industry. Examples of these kinds of revisions are the soft conversion to the metric system and editorial changes to improve sentence structure.

This manuscript was prepared from the *"Preprint" of the Proposed Amendments for the 1981 National Electrical Code®* (National Electrical Code Technical Committee Report) and the *National Electrical Code Technical Committee Documentation,* together with actions taken at the NFPA annual meeting in Boston on May 19–22, 1980, so that its publication date would coincide as closely as possible with publication of the 1981 Edition of the National Electrical Code®.

All explanations and interpretations are the personal opinion of the author and have no official or legal status. As always, the final decision on interpretation of any requirements is the sole responsibility of the authority having jurisdiction. Requests for Official Interpretations of any of the rules should be addressed to the National Fire Protection Association.

George W. Flach

Article 90
Introduction

SECTION 90-2. Scope. In Part (a), subparts (1), (2), and (3) are revised by adding the words "Installations of . . ." at the beginning of each sentence to point out that the rules in the National Electrical Code® (NEC) deal primarily with installation practices and do not contain sufficient information to properly judge the safety of conductors and electrical equipment. Other Sections require approval of materials and equipment by the authority having jurisdiction, who may depend on "listings" by qualified testing laboratories.

A new Part B is added to Article 555, covering Floating Dwelling Units. To correlate this revision in the Scope, this type of occupancy is added in Section 90-2(a). Since watercraft are not covered in Part (b), the words "other than floating dwelling units" are added to emphasize that floating dwelling units are the only type of watercraft covered by the NEC.

A Fine Print Note is added to Section 90-2(b)(5) to clarify the exemption of electric utilities. The intent has always been to exempt only that part of utility company operations directly related to metering, generating, transformation, transmission, and distribution of electric energy. It was never intended that the exemption apply to office buildings, garages, warehouses, and other buildings or premises not directly involved in generation, distribution, control, etc., of electricity. All of these structures must be wired in accordance with applicable rules in the Code.

SECTION 90-3. Since this section says that the Code is divided into nine Chapters and makes no mention of the fact that the Introduction is an enforceable part of the document, there has been confusion as to whether the requirements of Article 90 can be enforced. To overcome this problem, Section 90-3 is revised to state: "This Code is divided into the Introduction and nine Chapters." With this revision there can be no doubt about the status of Article 90. It is now definitely a legal part of the NEC.

SECTION 90-6. Another change which points out that the NEC does not contain enough information to properly evaluate materials and equipment is found in Section 90-6. The words "For specific items of equipment and material *covered* by this Code " are replaced by "For specific items of equipment and materials *referred to in* this Code."

SECTION 90-8. The NFPA Standards Council requested that metric equivalents be included in the NEC. The results of the Technical Subcommittee work are found throughout the text, and new Section 90-8 explains the methods used by the committee in making the soft conversion.

The modern form of the metric system is known as the International System of Units (SI). Values of linear measurement are followed by the approximate SI units in parentheses to obtain a soft conversion from one system of measurement to the other.

ONE
General

Article 100
Definitions

Definitions for **Fixed, Portable, and Stationary Appliances** are deleted. The Technical Subcommittee on Definitions of Stationary Appliances made this proposal as a revision for the 1978 Edition of the NEC, but it was held on the docket to allow sufficient time for other Code Making Panels to effect necessary changes by substituting other language for these terms.

In the "Findings" portion of the TSC report it was noted that these definitions had different meanings in various parts of the Code. In some cases the method of connection of the appliance to the branch circuit was governed by whether the authority having jurisdiction classified the appliance as fixed, portable, or stationary. With this change, the confusion created by these definitions should disappear.

Approved for the Purpose. Another Technical Subcommitee had the responsibility of reviewing the definition and use of the term "approved for the purpose." In its review, the committee found that the term was often used as a "catchall" where no special material or equipment other than general-use type was required. For example, in the 1978 NEC one of the wiring methods that is suitable for use in boatyards and marinas is listed in Section 555-6 as "rigid nonmetallic conduit approved for the purpose." This statement could have created a controversy between the inspection authority and the installing electrical contractor should the inspector have interpreted this rule to mean that rigid nonmetallic conduit had to bear special markings to indicate that it was suitable for use in boatyards and marinas. Here is an example of the use of the term where it had no real significance and only created interpretation problems for the enforcing authority.

The committee also noted that "approved for the purpose" was sometimes used to allow the authority having jurisdiction to accept new products that may appear in the marketplace between editions of the NEC. However, if this is the intent, the committee recommends that other wording be used to convey this message to the authority having jurisdiction.

As a result, the committee recommended deletion of the definition "approved for the purpose" and added a new definition for "identified" (as applied to equipment). All Code Making Panels were given a copy of the TSC report with instructions and to review the use of the term "approved for the purpose" and replace with "identified" if special conditions such as environment, more reliable performance, special construction, etc., are necessary.

Circuit Breakers. The Fine Print Note following the definition of a circuit breaker is expanded to more clearly indicate that circuit breakers without integral trip units are indeed included in the definition.

Dust-Tight, Rainproof, Raintight and Watertight. These definitions are modified by the addition of the phrase "under specified test conditions" to alert users of the NEC to the fact that testing is conducted under standardized conditions. In unusual applications the appropriate ANSI or NEMA Standard should be consulted to determine if the equipment will perform safely in the expected environment.

Ground-Fault Circuit-Interrupter. The definition of a ground-fault circuit-interrupter is revised so that this term now applies to personnel protection only. It also points out that the device must trip within a certain period of time whenever the ground-fault current exceeds a predetermined value.

Identified. This term was previously used to describe the grounded circuit conductor. It is now used to point out that special equipment is needed in some applications. This example appears in Section 700-6: "Transfer equipment shall be automatic and 'identified' for emergency use...."

Insight From. This is a new definition for phrases such as "within sight of," "insight from," "out of sight," "not in sight from," etc., which are used throughout the Code. The only place where an explanation of the meaning of "in sight from" appeared in previous editions of the NEC was in Section 430-4, and since this, or similar expressions are used in more than two Articles, a definition rightfully belongs in Article 100.

Section 110-12

Interrupting Rating. In the 1978 Edition of the NEC the meaning of "interrupting rating" is explained in Section 110-9. However, this term is also used in Article 240 and in accordance with the Scope of Article 100, a definition belongs here. The words "under specified test conditions" are added to alert the designer to the fact that the testing parameters may have to be reviewed before selecting a particular product.

Labeled and Listed. The National Fire Protection Association now has standard definitions for "listed" and "labeled," and the revisions reflect this fact.

Raceway. The word "enclosed" is added to make it clear that cable trays are not raceways. The previous definition did not include this word and resulted in some controversy about the status of channel-type cable trays.

Although the intent has always been that cable trays are only a supporting system and not a recognized wiring method, the definition of a raceway contained a loophole which created some doubt as to whether an open-channel cable tray was a raceway or not.

Service-Entrance Conductors, Sub-sets. This phrase is deleted because it is no longer used in Article 230—Services. Although a set of service-entrance conductors (a sub-set) is still permitted to be tapped from the main service conductors, calling these conductors a sub-set of service-entrance conductors does not improve clarity but instead causes confusion.

Article 110
Requirements for Electrical Installations

SECTION 110-2. In this Section the Fine Print Note is changed to include a reference to Section 90-6, and "identified" replaces "approved for the purpose" to conform with the changes made in the definitions as recommended by the Technical Subcommittee—Approved for the Purpose.

SECTION 110-3. The second sentence in Part (a)(1) is added to require descriptive marking on electric materials and equipment where testing indicates suitability for a specific purpose, environment, or application. This revision will certainly help everyone in the electrical industry in the selection of the right product for special and unusual conditions.

SECTION 110-12. The second paragraph requires closure of all unused openings in electrical equipment with materials that are equivalent in thickness and strength to the wall of the enclosure.

SECTION 110-16. In an effort to standardize numerical values of voltages throughout the Code, the Correlating Committee of the National Electrical Code® appointed a Nominal Voltage Technical Subcommittee. Although some of the suggestions made by the TSC were not accepted by various Code Making Panels, many members agreed that the word "nominal" should appear after a voltage level is mentioned. This has been done throughout the text and the first examples appear in the title of Section 110-16; following "600 volts" in Part (a); and the heading of the first column in Table 110-16(a), which now reads: "Voltage to Ground, Nominal."

Exception No. 2 to Part (a) is revised so as not to require working clearances as shown in the Table, where uninsulated live parts operate at a voltage not greater than 30 volts RMS or 42 V dc. Notice that there is no mention of power-limited (Class 2) circuits in the Exception. Therefore, the Exception applies to all low-voltage circuits regardless of the amount or value of current available.

SECTION 110-17. The sentence structure in Part (a) is improved by removing the redundant phrase "approved cabinets or other forms of." Approved enclosures are required to guard against accidental contact with live parts, and cabinets are only one type of enclosure. Since "enclosure" is all-inclusive, there is no need to mention cabinets.

SECTION 110-31. The last sentence in Part (c) is revised to require only locked doors and covers on metal-enclosed equipment (over 600 volts, nominal) accessible to the general public.

This change removes the requirement for locking doors on electrical equipment in manufacturing plants and similar occupancies, provided that warning signs are installed and only qualified personnel have authority to open doors or remove covers.

SECTION 110-34. The second paragraph of Part (c) is new and requires high-voltage warning signs at entrances to areas having exposed live parts or exposed conductors operating at over 600 volts, nominal.

TWO

Wiring Design and Protection

Article 200
Use and Identification
of Grounded Conductors

SECTION 200-2. To conform with the requirements in Section 250-152, this Section is revised to state that the neutral conductor insulation does not need to be equal to the insulation level of the ungrounded conductors for circuits operating at or above 1000 volts.

In other words, a neutral conductor with 600-volt insulation can be used on solidly grounded neutral systems rated 1000 volts or more.

SECTION 200-10. To allow more latitude in the design of attachment plugs, receptacles, and cord connectors, revisions are made in Part (b) to permit white marking at terminals rather than restricting identification to the use of the word "white" only.

The metal plating on wire-binding screw terminals is not required to be substantially white in color if the terminal for the grounded conductor is identified by a white marking or the word "white" appears adjacent to it. This change permits manufacturers of these devices to plate all terminals and screws with the same metal where corrosive atmospheres create unusual problems.

The Exception to Part (b) does not require terminal identification on two-wire nonpolarized caps. Since polarity is meaningless on two-wire nonpolarized caps, but is necessary if a polarized cap is to maintain continuity of the grounded conductor from the supply to an appliance or the screw shells of floor and table lamps, "nonpolarized" is added to the Exception to make it clear that identification of the grounded conductor terminal on all polarized plugs is mandatory.

Finally, the last revision in this Section appears in Part (e). Appliances having single-pole switches, single-pole overcurrent protective devices, or line-connected screw shell lampholders must have the grounded conductor terminal clearly identified so that a qualified installer will not connect the "white" wire to the wrong terminal.

Should the connections be reversed on a 120-volt appliance, unit switches, overcurrent devices, and the center contact on screw shell lampholders would be connected to the grounded circuit conductor. Under some types of internal failures, switches and overcurrent protection would be ineffective. Also, there is a greater chance of receiving an electric shock during lamp replacement because of reversed polarity.

Article 210
Branch Circuits

SECTION 210-1. Since electrolytic cells, cell line conductors, and wiring of auxiliary equipment in the cell line working zone are not subject to the requirements of Article 210, an Exception is added.

SECTION 210-4. For the last few years there has been much discussion about the practice of supplying split-wired duplex receptacles from a 120/240-volt three-wire branch circuit. Some enforcement agencies have not accepted this installation technique and have also refused to allow removal of the break-off tab so that one-half of a duplex receptacle could be controlled by a wall switch, even though both halves are connected to the same two-wire branch circuit. These various interpretations should now cease because of the new third paragraph.

When reading the new rule, notice that the requirement for simultaneous disconnection of all ungrounded conductors at the panelboard applies to dwelling units only. I assume that a two-pole circuit breaker or two single units connected together by a handle tie are considered necessary to reduce the possibility of injury to an electrician or homeowner making repairs or replacing a receptacle. Since normal procedure is to check for power in only one-half of a duplex receptacle and then turn that circuit off, an individual could get hurt when removing the device from the outlet box if the other half of the receptacle is still energized. Simultaneous disconnection of all ungrounded conductors supplying the receptacle removes this potential hazard.

SECTION 210-6. In Part (b) the voltage between conductors that supply lighting on poles, tunnels, and similar structures is increased from 500 to 600.

A Fine Print Note that refers to the definition of a receptacle is added at the end of Part (c)(1) to confirm the fact that split-wired receptacles are acceptable.

SECTION 210-8. With adoption of the 1978 Edition of the NEC, inspection agencies began requiring ground-fault circuit-interrupter protection for personnel on all receptacles installed in garages of dwelling units. Shortly after occupancy of their new residences, homeowners began complaining about spoiled food in freezers located in the garage, inoperative electric garage door openers, and other failures of appliances installed in the garage. To alleviate these problems, many GFCIs were either bypassed or removed.

Since clothes washers, gas clothes dryers, central vacuum cleaners, food freezers, garage door openers, and fluorescent fixtures can all be cord-and-plug-connected to 15 - or 20-ampere receptacles, and these types of equipment are often located in some area of the garage, many electrical inspectors adopted local amendments to exempt receptacles serving specific appliances and equipment from the requirement for GFCI protection. Others made attempts to process Tentative Interim Amendments (TIAs) which would not require GFCI protection for all receptacles in the garage, but they were not successful. One was finally accepted in 1980.

The result of a number of proposals to revise the requirement for ground-fault protection on all receptacles in a garage appears in subpart (2) of Section 210-8(a). Now at least one (not all) receptacle in the garage must have ground-fault circuit-interrupter protection for personnel. If additional receptacles are installed for the convenience of the homeowner, they must also be protected by a GFCI. Any other receptacles that are provided for specific appliances do not have to be GFCI-protected. Also, any receptacles that are not readily accessible (8 feet above the floor) do not require GFCI protection.

SECTION 210-19. The minimum ampacity of branch-circuit conductors for ranges of 8¾ kW or higher rating is 40 amperes. Previous editions listed No. 8 as the minimum wire size without stating whether the conductor material was copper or aluminum. This change is primarily for clarification.

SECTION 210-23. The 25-ampere circuit is deleted from the title and text because 25-ampere circuits are not recognized in Section 210-3 or listed in Table 210-24. However, this change should not be interpreted to mean that a 25-ampere circuit is not allowed to supply an individual load. Although a 25-

ampere circuit is no longer recognized as suitable for supplying multioutlet branch circuits, it is permissible to supply an individual appliance such as an electric water heater rated 4500 watts at 230 volts with No. 10 aluminum or copper-clad aluminum conductors protected by 25-ampere branch-circuit overcurrent divices.

SECTION 210-52. Requirements for the installation of receptacle outlets formerly appeared in Section 210-25. A Code Making Panel Subcommittee has editorially rewritten the text for clarification and improved continuity.

At least one outdoor receptacle must be installed for one- and two-family dwellings. Formerly, the rule applied only to one-family residences.

Article 215
Feeders

SECTION 215-1. An Exception is added to point out that the requirements for feeders do not apply to electrolytic cell rooms.

SECTION 215-2. Increased ampacities of certain sizes of conductors for a three-wire single-phase residential service are permitted by Note 3 to Tables 310-16 through 310-19. The Note also recognizes increased ampere ratings for feeders that supply the total load of the service for residential occupancies, whereas old Section 215-2 allowed only increased ampere ratings for feeders in one-family dwellings. To correct this obvious conflict, Section 215-2 and Note 3 to Tables 310-16 through 310-19 are revised to state that increased ampere ratings apply to both service conductors and feeders supplying dwelling units.

Part (a) specifies the minimum ampacity of feeder conductors based on the number of branch circuits connected to the feeder. Formerly, minimum wire sizes were mentioned; now ampacities are given. No. 10 wire is replaced with 30 amperes in Part (a) and 55 amperes replaces No. 6 in Part (b). To clarify the intent of these requirements without mentioning the conductor material (copper or aluminum), it is better to specify ampacity than one wire size for copper and another for aluminum. This change allows the designer to select the most economical conductor material and type of insulation, rather than restricting him to specific wire sizes.

Part (c) of this Section generally stated that feeder conductors had to be increased in ampacity if they could become overloaded at some time in the future. Because it was an unenforceable requirement, this Part has been deleted.

SECTION 215-6. To clarify the intent, "equipment" is added in front of "grounding conductors" in two places and a reference is made to Section

Table 220-2(b)

250-57. It is now clear that a feeder must include an equipment-grounding conductor, or the wiring method (metal raceway, armor, metallic sheath, etc.) must provide equipment grounding for utilization equipment connected to branch circuits in a panelboard supplied from a feeder.

SECTION 215-9. Although "approved for the purpose" has been deleted, only "listed" ground-fault circuit-interrupters should be used on feeders to provide personnel protection for 15- and 20-ampere receptacle circuits as specified in Section 210-8. This is the Section that permits GFCI protection on feeders in lieu of individual branch-circuit or receptacle protection.

Article 220
Branch-Circuit
and Feeder Calculations

SECTION 220-1. An exception is added to alert users to the fact that electrolytic cell rooms are not covered by this Article.

TABLE 220-2(b). The unit load per square foot for banks and office buildings is reduced from 5 to 3.5 watts. The double-asterisk note requires that an additional 1 watt per square foot be added for general-purpose receptacles.

With the emphasis being placed in energy conservation, modern office buildings are being designed with lower lighting levels, "task" illumination, and more efficient light sources. In some cases this has resulted in lighting loads of less than 2 watts per square foot. Requiring service equipment, transformers, feeders, and distribution panels based on 5 watts per square foot is wasteful of materials and increases construction costs unnecessarily.

Where lighting and receptacles are supplied at the same voltage, feeder calculations are based on 4.5 watts per square foot. Where the lighting is supplied at one voltage—say 480Y/277—feeder sizes, distribution panels, etc., are calculated at 3.5 watts per square foot and 120-volt receptacle loads are calculated at 1 watt per square foot.

Item 3 in Part (c) of Section 220-2 is added to allow a different method for calculating loads. Task lighting is often supplied with modular office furniture, and to require a unit-load calculation of 180 volt-amperes for each outlet results in oversize services, feeders, etc. This requirement also limits the number of outlets on a branch circuit.

With this change the actual volt-ampere ratings of the lighting fixtures supplied by outlets determine the load calculations, rather than the arbitrary figure of 180 volt-amperes for each outlet.

SECTION 220-3. The two or more 20-ampere small appliance branch circuits required in the kitchen, pantry, breakfast room, and dining room do not have to extend into the family room. Formerly, all receptacles in the family room had to be supplied from the 20-ampere small appliance branch circuits. This is no longer necessary because "family room" is no longer listed in Part(b).

The change in Part (d) correlates with the reduction in unit loading on a watts per square foot basis for office buildings and banks while also conserving materials. Notice that the calculated load must be used for sizing feeders and panelboards, but additional spare overcurrent devices and branch-circuit wires do not have to be installed where the connected load is less than the calculated load. In other words, overcurrent devices and branch circuits have to be provided only for the connected load, not for the calculated load.

SECTION 220-20. The additional words at the end of the sentence prevent the installation of a feeder for commercial cooking equipment which could be smaller than a branch circuit supplying a single appliance. Let's assume four kitchen appliances connected to the same feeder. For example, a 50-kW deep fryer, a 1.5-kW gravy heater, and two 250-watt infrared food warmers result in a feeder demand of only (52kW × 0.8) 41.6kW. This shows that the feeder ampacity could be less than the ampacity of the branch circuit supplying the fryer. The revision no longer permits this inconsistency.

SECTION 220-30 and **TABLE 220-30.** The format is changed for easier understanding and application.

Supplementary heating that is an integral part of a heat pump is figured at 65 percent, the same as central electric space heating. Since supplementary heat is often energized when the heat pump is running, the supplemental heat must be included in the calculations at 65 percent, together with the equipment kVA rating of the heat pump. In this case, the heating and air conditioning are coincident loads, and both must be included in the calculations. Or stated another way: the non-coincident-load rule mentioned in Section 220-21, which permits omission of the smaller of two dissimilar loads, does not apply.

To make it clear that the demands for air conditioning and electric space heating are not subject to additional derating, the word "plus" is added before "First 10 kW of all other load."

Voltage values are changed in this Section and Section 220-31 to conform with present distribution voltages.

SECTION 220-32. The Exception to Part(a)(2) permits calculation of feeders and services for multifamily dwellings without electric cooking by adding an assumed load of 8kW. The language has been corrected to point out

Table 220-2(b) 15

that the demand load should first be determined by the method outlined in Part B of the Article. After this, calculations can be made by using the Exception, and the smaller of the two should be used for sizing the feeder and service conductors. Usually, the method permitted by Section 220-32 will result in smaller services and feeders, but it is not a bad idea to calculate the load both ways.

SECTION 220-34. The demand factors for schools are intended to apply to the complete complex, including classrooms, offices, teachers' lounges, auditorium, gymnasium, etc. Table 220-34 should not be used for calculating the feeder size for portable classroom buildings because the calculated load and resulting feeder could be much smaller than required. For example, a single classroom building containing lighting and air conditioning could be supplied by a feeder having less ampacity than the ampere rating of the air-conditioning equipment if the demand factors permitted by the Table are used. The new third paragraph prevents this from happening.

SECTION 220-35. This is a new Section which allows additional loads to be connected to existing feeders and services provided that maximum demand in kVA is available for at least 1 year, the existing demand multiplied by 1.25 plus the added load does not exceed the ampere rating of the feeder or service, and overcurrent protection satisfies the requirements of Sections 240-3 and 230-90.

In many office buildings and manufacturing plants, demand meters are installed by electric utilities to adjust rates. These maximum demands may now be used as a means for determining whether the service and feeders have to be increased when additional electrical equipment is added.

The easiest way to handle the arithmetic is to multiply the maximum kVA by 1000 and divide by system voltage. Multiply the result by 1.25 and subtract this value from the ampere rating of the (feeder or service) conductors or overcurrent protective devices, whichever is smaller. The remainder is the load in amperes that may be added.

Article 225
Outside Branch Circuits
and Feeders

SECTION 225-1. The new Exception points out that electrolytic cell rooms do not have to comply with this Article.

SECTIONS 225-6, 225-10, 225-14, 225-18, 225-19. The Technical Subcommittee on Nominal Voltage recommended that the word "nominal" be added wherever "600 volts" appears in these Sections.

SECTION 225-6. Minimum conductor sizes for copper and aluminum used for open overhead spans are now specified. Previous editions of the NEC mentioned minimum sizes but did not state whether the sizes applied to copper or aluminum. These changes appear in subparts (1) and (2) of Part (a).

SECTION 225-10. Multiconductor cable (approved for the purpose) may be used as outside wiring on surfaces of buildings. The phrase in parentheses does not appear in the 1981 Edition of the NEC because only wiring methods judged suitable for this purpose as mentioned in Chapter 3 can be used for this application. Or put another way, multiconductor cables installed on the exterior of structures must be one of the types listed in Chapter 3 as being suitable for use outdoors.

Rigid nonmetallic conduit is added to the list of wiring materials that may be installed on the exterior of buildings. Since all rigid nonmetallic conduit that is listed or labeled by a recognized testing laboratory is sunlight-resistant, there is no reason to prevent use of the product outdoors. However, because of its coefficient of expansion, manufacturers instructions must be followed during installation to prevent pullout at joints or buckling caused by expansion and contraction during temperature and weather extremes.

SECTION 225-18. Clearances above ground for open overhead wiring are revised to agree with the National Electrical Safety Code ANSI-C 2.

Where the supply voltage does not exceed 150 volts to ground and the conductors do not pass over driveways, the minimum clearance is 10 feet. Conductors over residential driveways must be at least 12 feet above the driveway where the voltage is limited to 300 volts to ground. A clearance of 18 feet above grade is required for conductors above streets, alleys, roads, driveways on other than residential property, and parking lots subject to truck traffic. Formerly, a height of only 15 feet was required above areas subject to truck traffic, but the increased height of vehicles has made this elevation too low.

SECTION 225-19. Vertical, diagonal, and horizontal clearances of conductors from buildings are also revised to require additional clear space around roofs and walls. Other structures, such as signs, chimneys, tanks, etc., are added, and clearances are specified. These revisions are necessary to conform with requirements of the National Electrical Safety Code and to reduce the possibility of accidental contact with "live" wires by individuals.

Article 230
Services

SECTION 230-2. Where more than one service supplies a building, a permanent plaque or directory must be placed at each service location indicating where the other services are and which part of the building is served.

If we assume that two services supply one building, a sign at one service should generally state: "This service supplies the North Half of this Building. Another service for the Sourth Half is located in Room 132 at East and 1st Streets." At the other service the sign should read: "This service supplies the South Half of this Building. Another service for the North Half is located in Room 101 at East and 2nd Streets."

The information on the plaque or directory should be as concise and specific as possible, but should include sufficient information so that either service can be located quickly without hesitation.

A Part of Exception No. 3 is deleted along with the definition of sub-sets. The portion removed allowed separate sets of service-entrance conductors tapped from one service drop or lateral, or the sub-sets could be tapped from main service-entrance conductors. Since all this detail and the definition did not add anything to the Exception, it was eliminated. A similar change occurs in Part (b), which was formerly an Exception.

SECTION 230-22. Individual conductors for overhead services must have extruded thermoplastic or thermosetting insulating material applied over the conductor metal; however, the grounded conductor in a multiconductor cable can be bare.

Insulating materials are now specified to reduce the shock hazard to anyone who may accidentally come in contact with overhead wires. This action is prompted by a report from the Consumer Products Safety Commission of 14 accidents associated with "covered" wires, nine of which resulted in electrocutions.

SECTION 230-24. "Nominal" is added after "600 volts" in accordance with the recommendations of the Technical Subcommittee—Nominal Voltages.

The wording of Exception No. 2 for reduced clearance of overhead service drop conductors where passing over the roof of a building is revised to make it clear that service drops cannot pass over more than 4 feet of roof. The new language states that not more than 4 feet of service drop conductors can pass over the roof overhang.

Clearances from the ground for service drop conductors are revised to more nearly agree with requirements of the National Electrical Safety Code.

A clearance of only 10 feet above finished grade is acceptable above areas accessible to pedestrians, or from any platform from which the conductors may be reached provided that the conductors are insulated, cabled together, and supported with an effectively grounded messenger. Also, the supply voltage cannot exceed 150 volts to ground. Notice that this 10-foot clearance does not apply to individual open conductors. Where single conductors are used for service drops, clearances above ground have to be increased.

Where the voltage is over 300 volts to ground but not more than 600 volts, the minimum clearance to ground is 15 feet under restricted conditions. Any other overhead service drops not qualifying for lower clearances must be at least 18 feet above grade.

SECTION 230-28. All raceway fittings which are a part of a service mast that supports service drop conductors must now be identfied for use with service masts. This change was recommended by the TSC, which reviewed the use of the phrase "approved for the purpose."

SECTION 230-30. In Parts c and d of the Exception, the words "approved cable assembly with a moisture- and fungus-resistant outer covering" are removed because underground service conductors containing an uninsulated grounded conductor must now be "identified for underground use." Any cable that is so marked has a moisture- and fungus-resistant outer jacket or covering; therefore, including this statement in the Exception is not necessary.

SECTION 230-40. Part (b) of this Section, which formerly permitted covered (not necessarily insulated) service-entrance conductors to enter buildings or other structures, is deleted. This change was initiated by an accident report from the Consumer Products Safety Commission.

The Exception allowing an uninsulated grounded conductor is similar to the one in Section 230-30 and contains the same revisions.

SECTION 230-46. This Section requires that service-entrance conductors be installed without splices; however, there are Exceptions. Since the first four Exceptions do not recognize splices in busways, Exception No. 5 is added to clarify the fact that a busway used for service entrance may include splices for joining sections, and other fittings necessary to terminate the busway at the point where it receives its supply and also at the switchboard.

SECTION 230-48. A sentence is added to point out that care should be exercised in selecting raceway sealing materials so that compounds or tapes used do not have a deleterious effect on the cable insulation, shield, semiconducting layer, etc.

SECTION 230-50. Individual open service-entrance conductors must now be installed at least 10 feet above grade. Formerly, the height was only 8 feet above grade.

SECTION 230-56. This is a new Section which requires distinctive marking of the phase conductor having the higher voltage-to-ground on four-wire delta-connected services. Although many inspection authorities and electric utility metering personnel have required some form of marking on the higher-voltage leg, there was no requirement in the NEC. Previous to this revision, special marking of the higher voltage-to-ground phase conductor applied only to feeders as covered by Section 215-8.

SECTION 230-70. The sentence structure is revised to make it clear that the service disconnecting means must be suitable for use as service equipment in addition to being suitable for prevailing conditions. The old wording could be interpreted to mean that the service disconnecting means either had to be suitable for use as service equipment or the prevailing conditions or both. The new language clears up this ambiguity.

SECTION 230-71(a). To correlate with the changes made in Exception No. 3 and deletion of the definition for sub-sets of service-entrance conductors in Section 230-2, the text in Part (a) of Section 230-71 is revised to indicate that up to six switches or circuit breakers may serve as the service disconnecting means for each service permitted by Section 230-2.

SECTION 230-72. Part (a) is revised by deletion of the reference to one service drop or lateral. Since the intent is to permit up to six disconnecting means for each service, including any reference to one service drop or service lateral only causes confusion.

Part (b) is also changed to make it clear that a fire pump service or emergency service may each have up to six disconnecting means in addition to those permitted for the normal service.

In Part (d), covering multiple-occupancy buildings, the requirement for grouping service disconnecting means where there is individual occupancy above the second floor is deleted with this revision and the changes made to Exception No. 3 in Section 230-2. The authority enforcing the code can grant special permission for more than one service in a multiple-occupancy building, and the service disconnecting means do not have to be grouped even though there is individual occupancy above the second floor.

The paragraph in Part (d) which allowed service conductors to be run to each occupancy where there was no individual occupancy above the second floor is also deleted. However, this should not be interpreted to mean that services cannot be run to each occupancy by special permission. But placing a plaque or directory at each service which explains where the other services are located and what each one serves could be a tremendous task if, for example, individual services are provided for each apartment in an apartment building. In such a situation the directory could become so massive as to make this type of installation impractical.

SECTION 230-82. This Section generally prohibits connection of equipment on the line side of the service disconnecting means, but there are Exceptions. In an effort to limit the types of equipment that may be connected ahead of the service disconnect, the word "fuses" is eliminated from Exception No. 1 and the term "cable limiters or other current-limiting devices" is substituted. In Exception No. 2, fuses and disconnecting means or circuit breakers suitable for use as service equipment where installed in meter pedestals are permitted. However, the meter pedestal must now be recognized as suitable for use as service equipment. Both of these changes provide additional safety by requiring disconnection of the service before changing overcurrent devices.

In Exception No. 4, surge-protective capacitors are permitted to be connected on the supply side of the service disconnecting means. The word "devices" replaces "capacitors" because there are many surge protectors that are not capacitor-type.

SECTION 230-83. This Section formerly spelled out requirements for emergency transfer equipment. The title is changed and "alternate source" replaces "emergency service" to expand the requirement to all standby power sources, whether used for emergency, legally required standby, or optional standby systems.

The Exception permitted parallel operation only where suitable automatic controls were provided. This is now changed to allow manual switching for paralleling the alternate source with the normal source where suitable controls and protective equipment assure that manual switching will not create a hazard.

SECTION 230-84. This Section allows one service to supply more than one building on the same property provided that each building has a disconnecting means. A Fine Print Note suggested that Section 230-72, Parts (c) and (d), be referred to for proper locations of the required disconnecting means. Since Fine Print Notes are advisory (not mandatory) and the Code should specify the location of the disconnecting means for each building or structure, the Fine Print Note is changed to regular type. With this revision it

Section 230-201

is now required that the disconnecting means at each building be located as stated in Section 230-72, Parts (c) and (d).

SECTION 230-90. This Section deals with overcurrent protection for services. Exception No. 4 is revised by deletion of the following: "A multiple-occupancy building having individual occupancy above the second floor shall have service equipment grouped in a common accessible location. The overcurrent protection shall consist of not more than six circuit breakers or six sets of fuses." This change correlates with the revisions made in Sections 230-2, 230-71, and 230-72. With this change service equipment may be installed in each occupancy of a multiple-occupancy building without regard as to the number of floors. For example, a four-story apartment building can be wired so that service equipment is installed in each occupancy. However, at least two other requirements make this arrangement impractical. First, the service disconnecting means and overcurrent protection must be located at a readily accessible point nearest to the entrance of the service conductors into the building. Second, a plaque must be provided at each service-entrance location listing all the other services in the building.

SECTION 230-91. The acceptable practice of allowing service overcurrent protective devices to be located at the outer end of the service entrance conductors is removed. Since safety is enhanced by having the service disconnected before replacing overcurrent devices, they should be an integral part of the service disconnecting means.

SECTION 230-95. Exception No. 2 is added at the end of Part (a) to exempt fire pumps from the requirement for ground-fault protection.

Fire pumps are either supplied from a separate service or from one of the mains on the normal service. Where the fire pump motor is 150 hp or larger at 480 volts, the main is at least 1,000 amperes.

To improve the reliability for this important equipment under fire conditions, ground-fault protection should not be required.

SECTION 230-200. The word "nominal" is added for services exceeding 600 volts. The intent here is to alert Code users to the fact that system voltages as high as 635 are not to be judged under Part K of Article 230, but should be governed by rules for services not exceeding 600 volts.

SECTION 230-201. An Exception is added to permit some latitude in determining whether the primary or secondary conductors are classified as service conductors where both the primary and secondary voltages are above 600 volts nominal.

This change will give electrical engineers more freedom in system design for large industrial complexes.

The language in Part (b) is revised to more clearly state that the primary conductors are the service conductors under all conditions not covered by Part (a).

SECTION 230-202. Busways are recognized as a wiring method and are listed under item 7 of Part (b).

In Part (g) the phrase "identified for use in wet locations" is substituted for "approved for the purpose."

SECTION 230-205. Part (b) is added to require an accessible disconnecting means for the service conductors at the point of connection between the facilities of the serving utility and the premises' wiring.

Article 240
Overcurrent Protection

SECTION 240-2. Since medical and dental X-ray equipment is covered in Article 517, and Article 660 covers all other X-ray units, Article 517 is added to the list of equipment having special rules for overcurrent protection.

SECTION 240-3. Exception No. 4 is rewritten to state simply that overcurrent protection for remote-control circuits must comply with Article 725. Formerly, remote-control wiring was considered as protected from overcurrent where the protection did not exceed 300 percent of the ampacity of the conductors. This was not exactly correct because Section 725-12 allows 20-ampere overcurrent protection for No. 18 fixture wire, which has an ampacity of only 6 amperes. This revision puts the requirement for overcurrent protection of remote control circuits in Article 725, where it belongs.

SECTION 240-8. This Section does not permit fuses or circuit breakers to be connected in parallel. The Exception recognizes parallel araangements where factory-assembled and approved as a unit. The former wording was "approved for the purpose."

SECTION 240-21. Tap conductors not over 25 feet long now have to be protected from physical damage by being enclosed in a raceway. This is a change that appears in part d of Exception No. 3.

Overcurrent protection of conductors at the point of supply is not required at generator terminals according to Exception No. 9. However, overcurrent protection should be provided as close to the generator output terminals as possible.

Section 240-83

Exception No. 10 allows a feeder tap to be 100 feet long under controlled conditions. Notice that this Exception applies only to high-bay manufacturing buildings having walls not less than 35 feet high. Also, qualified electricians must be employed to maintain and repair the electrical installation.

Since the tap cannot run more than 25 feet horizontally, the full benefit of the 100-foot run will not be realized unless the side walls of the manufacturing plant are about 75 feet high.

Parts a, b, and c of the Exception are the same for the 25-foot tap rule, while Parts d through f are additional restrictions. These are: not allowing any splices; limiting the wire size to No. 6 copper, No. 4 aluminum, or larger; and not permitting the tap conductors to penetrate any walls, floors, or ceilings.

This change is quite controversial and generated negative comments from Code Making Panel members. There were also objections to acceptance of this revision from individuals during the public review period, but the change is based on proposals submitted for revision of the 1975 NEC wherein the submitter points out that this length of tap conductor has been successfully used for years in industrial plants with high ceilings.

SECTION 240-24. Although this Section requires overcurrent protective devices to be readily accessible, there are Exceptions. To correlate the action taken in Sections 230-82 and 230-91 by generally requiring service overcurrent protective devices to be an integral part of the service disconnecting means, reference to Section 230-91 is dropped from the first Exception.

An important change that will affect residential construction in various parts of the country is found in Part (d). Distribution panels and panelboards can no longer be located in clothes closets or similar areas containing easily ignitible materials. Other examples are towel and linen closets.

SECTION 240-40. To discourage the practice of installing fuses at the outer end of the service-entrance conductors as formerly permitted by Section 230-91, the language of Exception No. 1 to Section 240-40 is revised by substituting current-limiting devices for fuses.

SECTION 240-61. The classifications of fuses according to their ampere ratings are for 600 volts, nominal or less. Therefore, the voltage level is added in the text and Exception No. 2. The revision in the Exception also alerts Code users to the fact that medium- and high-voltage fuses should not be used on circuits operating at 600 volts or less.

SECTION 240-83. Where circuit breakers are used as switches for 120-volt fluorescent lighting, they now have to be approved for such switching duty instead of having to be "approved for the purpose." Circuit breakers tested for these conditions are marked "SWD."

Article 250
Grounding

SECTION 250-1. This Article covers grounding and bonding of electrical installations with specific rules itemized in Parts (a) through (f).

Part (g) has been removed as well as Article 250, Part M. Both of these references dealt with connections for lightning arresters. A Subcommittee of Code Making Panel 5 was appointed to review existing requirements relative to installation, connection, and grounding of lightning arresters, to expand the Article to include rules for installations at additional locations as well as industrial plants, and to update the language to agree with other standards. One such change appears in the title of rewritten Article 280; "Surge Arresters" replaces "Lightning Arresters." Another improvement is the consolidation of all requirements in one Article.

SECTION 250-7. Titles are added to Parts (a) and (b). This type of editorial revision is made in many places throughout Article 250 but will not be mentioned each time.

SECTION 250-23. A number of changes are made to Part (a) to clarify grounding of ac systems. The title of the Section and Part (a) apply to grounding connections from a service-supplied ac system and the changes reflect this fact. There is now distinct language to separate requirements for service-supplied systems from separately derived systems. Changes in these two Sections should reduce confusion. Grounding of ac services is covered by Section 250-23 and grounding of separately derived ac systems by Section 250-26.

For service-supplied ac systems, the grounding electrode must be one or more of the types specified in Part H of Article 250. These include metal underground water pipes, the metal frame of a building, a concrete-encased electrode, and a ground ring. Where all are available on the premises served, they must be bonded together to form a grounding electrode system. Other grounding electrodes mentioned in Section 250-83 may be used where none of the previously listed electrodes is available.

It is now clear that the grounding electrode conductor has to be connected to the grounded service conductor. This connection may be made at any accessible point between the load side of the service drop or lateral and the terminal or bus in the service disconnecting means on which the grounded service conductor is connected.

Where the service is supplied from an outside transformer, one additional grounding electrode and grounding electrode conductor must be connected to the grounded service conductor, either at the transformer or elsewhere outside the building. Notice that two or more grounds are required for a service-supplied system, whereas only one ground is necessary for a separately derived system. The revised wording in Exception No. 1 points this out.

In large switchboards, an equipment grounding bus for bonding sections together is usually provided. A bonding jumper in the service compartment bonds the grounded service conductor (bus) to the ground bus. According to Part (a), the grounding electrode conductor must be connected to the grounded service bus. However, Exception No. 5 now permits this connection on the equipment grounding bus. It is necessary to allow the grounding electrode conductor connection on this bus where a ground-fault protection system employs a ground-return sensor. With this type of ground-fault protection, the ground-return current from a phase-to-ground fault is detected at the bonding jumper. To assure proper functioning of this type of ground-fault protection, it is essential that the grounding electrode conductor be connected to the equipment-grounding bus, not the neutral bus.

Part (b) requires that a grounded conductor be brought into the service equipment where the voltage is less than 1000 volts and the supply is grounded. This conductor cannot be smaller than that shown in Table 250-94. If the phase conductors are larger than 1100 MCM copper or 1750 MCM aluminum, the grounded conductor must be not smaller than 12 ½ percent of the area of the largest phase conductor. The equivalent size for aluminum is new.

SECTION 250-25. This section specifies which conductor has to be grounded for ac premises wiring systems.

Since the grounded conductor has previously been referred to as the "identified" conductor, the words describing this conductor had to be changed when the Technical Subcommittee—Approved for the Purpose recommended that "identified" be used instead of "approved." This change will cause some confusion until everyone gets used to the new terminology. Regardless of the language used to describe the grounded conductor, it still must have a white or natural gray finish as required by Section 200-6.

SECTION 250-26. Because grounding methods for separately derived systems are not the same as those for service-supplied ac systems, a different set of rules apply. The major differences and revisions are discussed so that you will have a better understanding of grounding procedures.

Part (a) requires a bonding jumper sized either according to Table 250-94 or not smaller than 12 ½ percent of the largest derived phase conductor where wire size is greater than 1100 MCM copper or 1750 MCM aluminum.

This bonding jumper is connected from the grounded conductor of the derived system to the enclosure or grounding bus.

Formerly, the connection to the grounded circuit conductor had to be made at the source. Now the connection can be made at any point between the source and the first disconnecting means or overcurrent device. In other words, the bonding connection is permitted at any accessible point on the derived system as long as it is ahead of the first overcurrent devices or disconnecting means. For example, where a transformer supplies a switchboard, the bonding jumper can be installed either at the transformer terminal or in the switchboard. And for double-ended switchboards supplied by two transformers with a secondary tie, one bonding jumper in the switchboard is adequate if connected to the tie point of the grounded conductors.

Only one grounding electrode conductor is required to ground the grounded conductor of the derived system. Notice this difference when compared to a service-supplied system, which requires grounding at two or more separate locations.

Preferred grounding electrodes for a derived system are the effectively grounded metal frame of the building or the nearest effectively grounded metal water pipe. Other grounding electrodes listed in Sections 250-81 or -83 are acceptable only where grounded structural steel or water pipes are not available. The intent here is to limit the length of the grounding electrode conductor as much as possible while assuring a good low-impedance ground.

The two Exceptions for bonding and grounding Class 1 systems derived from transformers rated not more than 1000 volt-amperes allow a bonding jumper to be sized not smaller than the derived phase conductors, provided that the bonding jumper is not smaller than No. 14 copper or No. 12 aluminum. Also, a grounding electrode is not required for grounding the derived grounded circuit conductor where a bonding connection is made between this conductor and the transformer case, and the case is grounded by one of the methods specified in Section 250-57.

SECTION 250-44. Throughout Article 250 grounding rules apply to equipment operating at either 1000 volts or less or above 1000 volts. Before the voltage was increased to 1000, 750 volts was the "break" point. However, when the voltage level was raised from 750 to 1000, some values were not corrected. One such change appears in Part (d); others appear in Sections 250-121, -123, and -124.

SECTION 250-45. "Classified" is added in the heading and text to conform with the title of Article 500—Hazardous (Classified) Locations. Actually, areas are classified because of the nature of the combustible materials and their presence in sufficient concentrations to cause ignition or explosion. Since these areas are usually not hazardous at all times, it is not proper to simply define them as hazardous locations.

Part (b) states that all cord-and-plug-connected equipment operating above 150 volts to ground shall be grounded. Exception No. 2 removes electrically heated appliances from this provision if special permission is obtained from the enforcing authority and all exposed metal is permanently and effectively insulated from ground. Since many electrical inspectors are very reluctant to grant special permission, this Exception will not receive wide acceptance.

Although there are many rules in the NEC that may be waived by obtaining special permission from the electrical inspector (authority having jurisdiction), he should not abuse this privilege. If after a careful and complete evaluation of all the facts and possible consequences, the inspector is convinced that a waiver is justified, only then should special permission for a deviation from the basic rule be granted.

SECTION 250-53. Since this Section deals with the grounding path between the grounding electrode and the grounded service conductor, the word "service" is added in the title and Part (a). The Fine Print Note is also added to direct the reader to Section 250-23(a), which contains more detailed information on connecting the grounding electrode conductor into the service equipment.

SECTION 250-57. Noncurrent-carrying metal parts of fixed equipment, raceways, and other enclosures must be grounded by one of the methods indicated in this Section. "Raceways and other enclosures" are added because proper grounding of these items is not spelled out in any other part of Article 250. However, this revision makes the Exception necessary to avoid a conflict, because Sections 250-60 and -61 recognize grounding of some equipment, appliances, raceways, etc., through the grounded (neutral) circuit conductor. Specifically, Section 250-60 permits grounding the frames of ranges, clothes dryers, and associated wiring components with the grounded conductor. Section 250-61 recognizes the grounded circuit conductor as an acceptable grounding means for service raceways, meter bases, and other equipment on the line or supply side of the service disconnecting means.

Part (c), which allowed other methods for grounding fixed equipment by special permission, is removed because Section 90-4 allows the authority having jurisdiction to waive specific requirements where equivalent safety can be assured. In other words, this part was redundant.

SECTION 250-59. In Part (c) the words "or by special permission" are removed because, again, they are redundant.

SECTION 250-61. In Part (a) the grounded circuit conductor can be used to ground all raceways, enclosures, and equipment on the line side of the service disconnecting means. Previously, service raceways and meter enclosures were mentioned as examples. This change makes it clear that all types of electrical components and equipment may be grounded in this manner.

SECTION 250-71. Effective August 1, 1981, services for dwellings have to be installed so that there is an accessible point (external to enclosures) for connecting intersystem bonding and grounding conductors.

Communication circuits, antenna protection systems, and CATV cable must be grounded, preferably to some part of the service grounding system. With the increasing use of plastic water pipe and nonmetallic conduit for service raceways in residential construction, an accessible ground for intersystem bonding is no longer available. To overcome this problem, an external lug, connector, or stud will have to be installed on the service equipment where there is no exposed metal service raceway, grounding electrode, or grounding electrode conductor. It is necessary to bond communication and CATV systems to the electric service to eliminate differences in potentials between the various systems caused by power faults or lightning. Improper interconnection of the systems can result in electric shock and fire hazards.

SECTION 250-72. There is continuing confusion about proper bonding of service equipment. Misunderstandings usually occur in the interpretation of Part (e), which deals with locknuts and bushings. Double locknuts, one inside and one outside an enclosure, are not acceptable.

Since this is one method that is recognized for bonding equipment operating above 250 volts and the wording in Part (e) was not clear to some Code users, another attempt is made to clean up the language so that only set-screw or clamping-type locknuts and bushings are used. Remember that bonding jumpers must also be used where the largest eccentric or concentric knockout is not removed from the enclosure.

SECTION 250-78. The revision more accurately describes the items of the electrical system that must be bonded in hazardous (classified) locations.

SECTION 250-81. A grounding electrode system consisting of metal underground water pipe, the metal frame of a building, a concrete-encased electrode, and a ground ring must all be interconnected to form a single grounding electrode if all or some of these types of grounding electrodes are available at the building or structure supplied by an electric service. These grounding electrodes must be connected to each other with a bonding jumper whose size is selected from Table 250-94. The grounding electrode conductor is sized in the same manner and may be connected to the most convenient grounding electrode in the system. This change makes it clear that the grounding electrode conductor does not have to terminate on the metal water pipe.

In many buildings, especially dwellings, the only grounding electrode available is a metal water pipe. Where this condition exists, an additional grounding electrode must be installed. This supplemental grounding electrode is usually a ground rod. Since Section 250-81(a) was not clear as to where and how this supplementary electrode should be connected into the grounding system, a sentence is added which allows bonding to any of the following: the metal water pipe, the grounding electrode conductor, the grounded service-entrance conductor, or the metal service raceway.

Part (c) formerly specified No. 4 solid copper wire for a concrete-encased electrode. This is changed to allow either a solid or a stranded copper conductor.

SECTION 250-83. Installation rules for ground rods are expanded in Part (c)(3). Eight-foot ground rods now have to be driven flush with the earth. Where a rock bottom is encountered, the rod has to be driven at an angle or buried horizontally in a trench at least 30 inches deep. If the rod is longer than 8 feet, that portion in excess of 8 feet can remain aboveground where not subject to physical damage.

Ground clamps that will be buried in the earth should be carefully selected, because those made of aluminum, steel, or similar metals might fail after a short period of time.

SECTION 250-84. Where a made electrode (rod, pipe, or plate) has a resistance of more than 25 ohms, an additional electrode must be provided. If this second electrode is placed close to the first, the ground resistance is only slightly lowered. To be effective in reducing the resistance to ground, the electrodes should be spaced at least 6 feet apart, and this is now required by the last sentence.

The Fine Print Note points out that the efficiency of rods in reducing earth resistance is improved by increasing the distance between them to more than 6 feet if they are more than 8 feet long.

SECTION 250-91. In the list of acceptable types of equipment-grounding conductors, flexible metal conduit and fittings are mentioned; however, both must be approved for grounding. All Underwriters' Laboratories, Inc. (UL)-listed fittings and conduit in the ⅜- through ¾-inch trade size are considered suitable for grounding provided that the contained conductors are protected by overcurrent devices rated 20 amperes or less and the length is not more than 6 feet. Flexible metal conduit and fittings larger than ¾ inch must be marked or otherwise identified to qualify as a grounding method.

SECTION 250-92. Number 4 or larger copper or aluminum grounding electrode conductors must be protected from severe physical damage. The reference to aluminum is new.

Where protection from physical damage is necessary, the conductor can now be enclosed in rigid nonmetallic conduit. This product has now regained recognition which it had in the 1975 NEC but did not have in the 1978 Edition.

The use of nonmetallic conduit as protection for the grounding electrode conductor should reduce the impedance to a value below that which would be obtained if the conductor were installed in a ferrous raceway.

The addition in the second paragraph simply states that nonmetallic conduit must be installed according to the rules in Article 347—Rigid Nonmetallic Conduit.

The reference to Section 210-7 is removed from paragraph (b)(2) because the Exception in this Section was deleted in the 1978 NEC. The Exception in Section 250-50 permits a separate equipment grounding conductor to a cold water pipe where a grounding-type receptacle is added to an existing branch circuit that does not include an equipment grounding conductor. Under all other conditions where equipment must be grounded, the raceway or cable must include or provide an equipment-grounding means.

SECTION 250-93. The size of the grounding conductor for dc systems is specified in Part (c) as not being smaller than No. 8 copper. An aluminum conductor not smaller than No. 6 is equally acceptable. The equivalent size for aluminum is also added in Section 250-94.

SECTION 250-95. This Section details procedures for proper sizing of equipment grounding conductors, but was vague where a common equipment ground was provided for more than one circuit in a raceway. Generally, the ampere ratings of all overcurrent devices supplying circuits connected to the same phase were added together and the equipment grounding conductor was selected from Table 250-95 for this total current. To show how this was done, let's assume we have six 20-ampere 120-volt circuits supplied from a 120/240-volt single-phase distribution panel. The circuits are installed in nonmetallic conduit with the equipment grounding conductors. Is it necessary to pull three No. 12 grounding conductors (one for each three-wire circuit), or will one suffice? If only one is necessary, what size should it be? Certainly, three No. 12 equipment grounding conductors will satisfy Code rules, but are they really necessary? Common practice has been to add up the ampere ratings of all overcurrent devices connected to the same phase. In this example the total is (3 × 20) 60 amperes, and according to Table 250-95, a No. 10 equipment grounding conductor could serve as a common ground for the six circuits. This method is based on the assumption that phase-to-

Section 250-124

ground faults will occur simultaneously on all circuits connected to the same phase—a highly unlikely condition.

The last paragraph in Section 250-95 now makes it clear that a single equipment grounding conductor can be used where more than one circuit is installed in the same raceway. This conductor's size is determined by the largest overcurrent device protecting any circuit installed therein.

TABLE 250-95. As you know, the size of equipment grounding conductors is based on the ampere rating of the overcurrent device protecting the circuit conductors and equipment. Since there were large gaps between 200 and 400 amperes and between 400 and 600 amperes, sizes of equipment grounding conductors for 300- and 500-ampere circuits are added. Formerly, No. 3 copper wire had to be used for the equipment ground on a 300-ampere circuit and No. 1 copper for a 500-ampere circuit. Now No. 4 and No. 2, respectively, are listed for these ampere ratings.

SECTION 250-98. This Section in the 1978 NEC was confusing and is deleted because it permitted grounding conductors to occupy the same raceway with other conductors, where in fact the grounding conductor is required to be run with the circuit conductors and in the same raceway by Sections 250-57, -91(b), -92(b), and -95.

SECTION 250-111. This Section in the 1978 NEC is also deleted because it referred to equipment grounding when Table 250-95 included pipe as a suitable grounding means. The column listing minimum pipe sizes for equipment grounding was eliminated in the 1965 NEC.

SECTION 250-115. Since Section 250-83(b)(3) states that rod and pipe electrodes must be installed so that 8 feet of length is in contact with the soil, ground clamps will necessarily be buried where the electrode is only 8 feet long. A revision in this Section requires that ground clamps used on pipe, rod, or other buried electrodes be suitable for direct burial in the earth. In other words, steel or aluminum clamps and those held together with steel screws should be avoided.

SECTIONS 250-121, 250-123, 250-124. These Sections provide rules for grounding or not grounding (under certain conditions), instrument transformers, meters, relays, etc. The requirements applied to voltages of 750 or less and over 750 volts. Because the rest of Article 250 separates the rules to electrical systems of less than 1000 volts with different requirements for over 1000 volts, the changes here are the substitution of 1000 volts for 750 volts, which formerly appeared. These changes make the voltage level consistent throughout Article 250.

SECTIONS 250-154, 250-155. These Sections deal with grounding of equipment operating at and above 1000 volts and connected to the supply through portable cables. The word "mobile" is added wherever "portable" appears because large mobile equipment, which is usually self-propelled, is not portable in the sense that it can be picked up and carried.

Article 280
Surge Arresters

In addition to the title change from "Lightning Arresters" to "Surge Arresters," the Article is completely rewritten. The material is updated and consolidated. In past editions of the NEC, specifications for connecting surge arresters (lightning arresters) were placed in Part M of Article 250.

One major change is that surge arresters are no longer required at industrial stations where thunderstorms are frequent. However, if arresters are installed, they must comply with Article 280.

SECTION 280-1. As seen in the Scope, the Article covers general requirements, installation requirements, and connection requirements for surge arresters.

SECTION 280-3. A single set of surge arresters is permitted to protect more than one interconnected circuit provided that no disconnected circuit is exposed to surges while not connected to the arresters.

SECTION 280-4. On circuits rated less than 1000 volts, the surge arrester rating must be equal to or greater than the maximum phase-to-ground power frequency voltage.

Arresters connected to systems operating at 1 kV and above must have a rating of not less than 125 percent of the maximum continuous phase-to-ground voltage. For example, a 4160Y/2400-volt grounded system requires surge arresters with a minimum voltage rating of (2400 × 1.25) kV.

SECTION 280-11. Surge arresters can be installed indoors or outdoors and must be inaccessible to unqualified persons unless listed for installation in accessible locations.

SECTION 280-12. Conductors used to connect surge arresters must be as short as possible without sharp bends.

SECTION 280-21. Leads for connecting surge arresters cannot be smaller than No. 14 copper or No. 12 aluminum on systems of 1 kV or less. Where installed at the service, the ground terminals of the arresters must be

Section 280-24

connected to any of the elements that are a part of the service-grounding system.

SECTION 280-22. Where the arrester is installed on the load side of a service of less than 1 kV, the arrester can be connected between any two conductors.

SECTION 280-23. Connections from the arrester have to be at least No. 6 copper or aluminum where used on circuits over 1 kV.

SECTION 280-24. The grounding conductor of a surge arrester connected to a system of 1 kV and above may be interconnected to the secondary neutral of a transformer that supplies a secondary distribution system where the secondary grounded conductor is connected to a continuous metal underground water-piping system. Where the distribution system is in an urban area with at least four water-pipe connections on the neutral in each mile, the arrester can be interconnected to the neutral and the direct grounding connection to the surge arrester can be omitted.

Interconnection between the grounding conductor of a surge arrester and the grounded conductor of a transformer secondary is also acceptable where the primary neutral has at least four grounds per mile, the secondary neutral is part of a multiground system, and there is a ground at the service.

THREE

Wiring Methods and Materials

Article 300
Wiring Methods

SECTION 300-1. This Section states that this Article applies to all wiring installations. Some individuals not completely familiar with the Code interpret this language to mean that building wire such as TW, THW, FEP, etc., can be installed as single conductors if protected from physical damage. To overcome this misunderstanding, Part (c) is added. Now we have a definite rule that prohibits single conductors unless installed as part of a recognized wiring method.

SECTION 300-5. Direct buried conductors emerging from the ground must be protected from physical damage. In the 1978 Edition of the Code raceways installed on poles to provide this protection had to be rigid metal conduit, intermediate metal conduit, or PVC schedule 80 conduit. The revision removes the reference to poles. This change can be interpreted as allowing schedule 40 PVC for protection of conductors emerging from the ground at poles or other locations where it has been determined that protection from physical damage is not necessary. However, before making this assumption, check with your electrical inspector.

SECTION 300-15. This Section generally requires a box at each outlet, switch, or receptacle location. However, Exception No. 5 to Part (b) permits the installation of self-contained (boxless) devices with integral enclosures in walls or ceilings of on-site construction where used with nonmetallic sheathed cable.

Originally designed for the mobile home industry, these devices with integral enclosures were accepted for use in on-site construction when the 1975 NEC was adopted. Since the Exception did not specify how these devices must be fastened in place, a change now appears which states that the device must have brackets that fasten it securely to a structural member.

Exception No. 6 is added to cover connectors and receptacles which are part of a prefabricated flexible wiring system. Although this method of wiring lighting fixtures and receptacles in office buildings has been in use for a few years, special treatment has not been given to manufactured wiring systems until now. New Article 604 covers construction and installation of these systems and will be discussed later.

SECTION 300-17. This Section limits the number and size of conductors in raceways. The Fine Print Note is updated to include new wiring methods and others that were not previously included.

SECTION 300-18. Old Part (b), which suggested that conductors not be inserted in raceways until the interior of the building was protected from the weather, is deleted because as written this paragraph was not enforceable.

SECTION 300-20. Skin-effect heating for pipelines and pressure vessels is accomplished by placing a single conductor in a ferromagnetic raceway. Since this Section requires ac circuits to be installed so that heating by induction is avoided, an Exception must be made for skin-effect heating systems.

SECTION 300-21. The title is changed to include products of combustion, and the body of the Section is rewritten to require firestopping with approved methods around openings for electrical penetrations through fire-rated walls, partitions, floors, and ceilings.

Underwriters Laboratories, Inc., lists a number of products for firestopping conduit and cable penetrations through fire-resistance-rated floors, walls, etc. Many of these devices serve a dual function by providing a firestop for the internal wiring while maintaining the fire rating around the hole cut into the fire-resistive building component.

SECTION 300-22. Type MC cable is an acceptable wiring method in ducts and plenums used for environmental air only if it does not have a nonmetallic outer jacket. This change appears in Parts (b) and (c).

Section 305-2   41

The Fine Print Note is added to more clearly define environmental ducts and plenums. Another Fine Print Note following Part (c) describes spaces used for supply or return of environmental air. With these explanations, more uniform interpretations should result.

The length of flexible metal tubing in environmental air ducts and plenums was limited to 4 feet. In hollow spaces used for movement of environmental air, a single length could not exceed 6 feet. Both of these restrictions are removed, and this wiring method can be installed as permitted by Article 349.

Another wiring method that is acceptable in the space used for environmental air is factory-assembled multiconductor control or power cable specifically listed for this application.

Manufactured metallic wiring systems covered by new Article 604 are acceptable in air-handling ceilings if listed for this use.

In residential construction it is common practice to use the space between wall studs and ceiling joists as a return air duct. Exception No. 5 permits nonmetallic cable to pass through these areas provided that the cable passes through perpendicular to the studs or joists.

Article 305
Temporary Wiring

SECTION 305-1. The language in Part (b) is cleaned up to make it clear that temporary power and lighting installations for Christmas decorations, carnivals, and similar purposes cannot remain for more than 90 days. Temporary wiring for experimental and development work was included in the 90-day limit but is now moved to a new Part (c). This Part is also expanded to include temporary wiring for emergencies and testing. There is no time limit on temporary wiring for these uses, but is must be removed promptly after it is no longer needed.

SECTION 305-2. Part (c) formerly required fastening of open conductors at ceiling height. This could not be accomplished outdoors, nor was it practical in high-bay buildings where the ceiling is 25 or more feet above the working level. For these reasons the reference to ceiling height is deleted. However, open wires still have to be located where they are not subject to physical damage and supported every 10 feet.

Part (c) also contained a statement that no conductors shall be laid on the floor. To make it clear that extension cords may lay on the floor, the wording is changed to state that no branch-circuit conductors shall be laid on the floor.

The second sentence in Part (d) is new and prohibits receptacles on lighting circuits. The intent of this change is to prevent loss of lighting.

Since all 15- and 20-ampere 120-volt receptacles connected to temporary circuits on construction sites must be protected by ground-fault circuit-interrupters, tripping of a circuit breaker-type GFCI supplying both lighting and receptacles will disconnect the lights and possibly create an unsafe condition.

Old Part (e), which prohibited bare conductors and earth returns, is deleted. The remaining Parts are relettered.

The second paragraph under Part (f) recognizes brass shell paper-lined sockets and other metal-enclosed sockets for construction-site lighting where the shells are properly grounded.

Duplicate rules for temporary wiring over 600 volts are removed together with subheadings "A—600 volts, nominal or less" and "B—over 600 volts, nominal." Old Section 305-11—Guarding is renumbered 305-5.

Article 310
Conductors
for General Wiring

SECTION 310-2. The first change in this Article is for clarification and appears in Part (b). Readers are alerted to the fact that conductors covered by the rules are copper, copper-clad aluminum, or aluminum unless other metals are specifically mentioned.

SECTION 310-4. Conductors smaller than No. 1/0 cannot be run in parallel. This is the general rule, but an Exception permits paralleling of small conductors in elevator traveling cables. Another Exception is added to permit conductors smaller than No. 1/0 in parallel for control power to indicating instruments, contactors, relays, etc., where the conductors are in the same raceway or cable, the ampacity of each conductor in the parallel arrangement is sufficient to carry all the load current by itself, and the overcurrent protection does not exceed the ampacity of one of the parallel conductors. Notice that paralleling conductors in these small sizes does not increase the ampacity of the combination. And overcurrent protection cannot exceed the ampacity of a single conductor.

The primary reasons for paralleling control wiring are to reduce voltage drop and capacitance between conductors.

SECTION 310-5. The minimum size of conductors for various voltage ratings is now shown in Table form. This and other changes integrate "under 600 volt" conductors with "over 600 volt" conductors. Headings "A—

General," "B—Conductors 600 volts, nominal or less," and "C—Conductors over 600 volts, nominal" have been removed as part of this editorial rearrangement.

Exception Nos. 8 and 9 allow wire sizes smaller than shown in the Table for various voltage ratings and insulation types. This is not new material, but it is necessary because of the conductor ranges shown in the new Table.

Because motor control circuit wiring can be as small as No. 18AWG, Exception No. 10 recognizes this fact.

SECTION 310-6. This is old Section 310-61 with "qualified testing laboratory" substituted for "nationally recognized testing laboratory."

SECTION 310-7. Instead of having the requirements for underground conductors spelled out in this Section, the reader is referred to Sections 300-5 and 710-3(b). The Section now applies to cables rated above 2000 volts. Notice that the shield, sheath, or armor must be grounded through a path that has sufficient capacity to conduct safely any fault current likely to be imposed on it.

SECTION 310-8. The negative statement "conductors shall not be used for direct burial in the earth unless approved for the purpose" is replaced with "conductors used for direct burial applications shall be of a type listed for such use."

SECTION 310-10. This Section generally states that conductors shall not be used in any manner that will result in exceeding the (insulation) temperature rating of the conductor. The Fine Print Note is new and contains much information that must be studied carefully. Notice that conductor temperature must be considered for the entire length of the conductor. For example, an underground feeder going from one building to another passes over hot water lines. Correction factors for the elevated ambient temperature near the water pipe must be used to obtain the maximum load current, or premature failure will result. I am aware of failures of wiring in conduits where hot water pipes and the raceway come within a few inches of each other in a concrete slab. When the conductors were pulled out of the conduit and laid on the slab above the approximate location of the conduit run, the brittle and charred insulation on the conductors matched the approximate location of the water pipe. A new conduit was run overhead to eliminate the problem. This experience shows the necessity for proper planning or derating of conductors where high ambient temperatures cannot be avoided.

Four items that determine the operating temperature of conductors are the ambient temperature, heat generated in the wire because of current flow, rate of dissipation of conductor generated heat, and proximity to other current-carrying conductors. Notes to the Ampacity Tables provide correction factors for these conditions and must be properly applied to ensure against premature insulation failure.

In residential wiring it is common practice to run many nonmetallic sheathed cables side by side on the same stud or ceiling joist. Often, one cable is placed on top of another and both are secured with a single staple. Since these cables are close to each other, overheating can result, especially if the cables are embedded in thermal insulation. Notice the language in item 4 of this Section. It cautions that adjacent load-carrying conductors—not necessarily in the same raceway—have the dual effect of raising the ambient temperature and impeding heat dissipation. Remember this discussion when wiring dwellings with NM cable.

SECTION 310-11. The marking (identification) of power-limited tray cable (PLTC) is permitted on the plastic jacket under the metal sheath.

SECTION 310-13. In consolidating "below 600 volt" and "above 600 volt" conductors, revisions are made to include all Construction and Application Tables under this Section.

SECTION 310-14. Only aluminum alloy is suitable for No. 8, 10, and 12 solid aluminum conductors.

To comply with revised UL Standards for aluminum wire and wiring devices, aluminum alloys were developed. At the same time, wiring device manufacturers redesigned switches and receptacles for use with this new alloy, and those found suitable bear the marking CO/ALR on the mounting yoke. Requirements elsewhere in the Code state that switches and receptacles that are directly connected to aluminum conductors must be marked CO/ALR. Therefore, aluminum alloy wire and CO/ALR devices must be used together to comply with the NEC.

SECTION 310-15. Old Section 310-38 is added here to consolidate information for all voltage ratings.

TABLE 310-13. The Note represented by the triple asterisk is new and recognizes a single conductor USE cable with 80 mils of insulation in sizes 213 to 500 MCM.

TABLES 310-16 and -17. The Technical Subcommittee that integrated the "under" and "over" 600 volt rules also revised the ampacities of conductors.

The ampacities of most conductors in the 85°C column are increased. Here are some examples using copper conductors: No. 2 increases from 120 to 125, No. 2/0 from 185 to 190, and 300 MCM from 300 to 310 amperes. Similarly, ampacities are increased for 90°C. Formerly, the ampacity of

90°C conductors was the same as 85° conductors. Now the ampacity of a 90° C conductor is about 4 percent greater. Although some of these changes are significant, they have not caused the controversy that has arisen because of increased ampacities for conductors in sizes Nos. 14, 12, and 10.

The ampacities for No. 14 and 12 are now listed in Table 310-16 at 20 and 25, respectively, in the 60°C and 75°C columns, and 25 and 30 in the 85°C and 90°C columns. Number 10 is also increased to 35 in the 75°C column. But these increases lose some of their glitter on closer examination. Notice the obelisk (some people call it a dagger) in front of the insulation types and in back of the ampacity figures; then read the corresponding footnote. Here we find that the load current and overcurrent protection cannot exceed 15 amperes for 14 AWG, 20 amperes for 12 AWG, and 30 amperes for 10 AWG copper. Aluminum and copper-clad aluminum are limited to 15 amperes for 12 AWG, and 25 amperes for 10 AWG. A logical question at this point is: What good are the increased ampacities? Let's explain with an example. Suppose that you are installing 16 No. 12 THHN copper conductors in ¾-inch conduit to supply 208Y/120-volt lighting and receptacle circuits. Also assume that the conduit passes through an area where the ambient temperature is 105°F. To find the allowable load current for each conductor under these conditions, we first apply a correction factor for the elevated ambient temperature.

Below the Ampacity Table is a list of Correction Factors. The statement at the top of this Table explains its use. Now we multiply the ampacity of No. 12 THHN copper (30 amperes) by the correction factor corresponding to 105°F, which is 0.87. This gives us an allowable load current of 26.1 amperes. But wait, we are not finished yet. A derating factor must be applied because there are 12 (neutrals carry unbalanced currents only) current-carrying conductors in the raceway. In accordance with Note 8 to Tables 310-16 through -19, the conductors must be derated to 70 percent of their adjusted ampacity. Multiplying 26.1 by 0.70 results in an allowable load current of 18.3 amperes. These conductors can be protected by 20-ampere overcurrent devices, and the 90°C temperature rating of the insulation will not be exceeded as long as the load current is not greater than the calculated value.

Another example using a continuous load will aid in making a comparison between the ampacity of No. 12 Type THW as listed in the 1978 Edition and as now revised in Table 310-16. Calculations are made by using the former ampacity of 20 and the new ampacity, which is 25. Assume that two 3-phase, 4-wire 480Y/277-volt branch circuits in a single raceway supply fluorescent lighting in an office building. Since the lighting is energized for more than 3 hours, a continuous load is involved and branch-circuit loading cannot exceed 80 percent. Also, according to Note 10 of Tables 310-16 through 310-19, the neutrals are current-carrying conductors, and derating is required by Note 8 of these Tables. There are eight current-carrying conductors in the conduit.

The ampacity of No. 12 THW copper conductors is given as 20 in Table 310-16 of the 1978 NEC. Therefore, the total load cannot exceed (0.70 × 20) 14 amperes. Applying the 70 percent derating factor required by Note 8 to the increased ampacity shown in Table 310-16 of the 1981 NEC results in a permissible load current of (0.70 × 25) 17.5 amperes. This load current (17.5 amperes) could be connected to the branch circuits if the panelboard assembly and overcurrent protective devices are listed for continuous operation at 100 percent of their rating. But this is not usually the case. Therefore, the branch-circuit load cannot exceed 16 amperes. Even with this reduction to 16 amperes, because the overcurrent protective devices cannot carry their 20-ampere rating continuously, there is still a gain of 2 amperes on each branch circuit when compared with the 14-ampere load permitted under the 1978 Code.

These examples show the usefulness of the increased ampacities for branch-circuit-size conductors, but for ordinary wiring where derating and continuous loads are not involved, remember the limitations placed on these small wire sizes by the footnote at the bottom of the Tables.

Note 3 to the Tables is revised to make it clear that increased ampere ratings for three-wire single-phase services and feeders apply to all dwelling units regardless of the number of dwelling units in a building.

Notes 8, 10, and 11 are changed to more clearly state that derating does not change the ampacity of the conductors in a raceway, but does reduce the maximum load current.

Data processing and similar electronic equipment produce third harmonic currents in three-phase four-wire wye systems. Therefore, they are added to Part (c) of Note 10, which means that the neutral of a circuit supplying this equipment must be considered as a current-carrying conductor when applying the derating factors listed in Note No. 8.

There are no revisions to the Ampacity Tables (310-39 through 54) for insulated cables rated 2001 volts and above.

Article 318
Cable Trays

SECTIONS 318-2, 318-8, and 318-9. Numerous conductor sizes are listed in these Sections, and some are not related to conductor ampacity. The revisions clarify the text by indicating that in certain cases wire sizes apply to both copper and aluminum conductors.

SECTION 318-12. The ampacity of medium-voltage multiconductor cables installed in trays is generally taken from Tables that list ampacities for three conductors in an isolated conduit in air. Exception No. 2 permits use of

Tables that are based on three-conductor cable in air where cables are installed in a single layer with a minimum spacing of one cable diameter between them, and the tray is not covered. This change is significant, as shown by a comparison of allowable ampacities for 250 MCM copper. Before this revision the ampacity of 5-kV cable was 280; now it is 320.

Article 320
Open Wiring on Insulators

SECTION 320-3. Open wiring on insulators is permitted only for industrial or agricultural establishments. The word "only" is added to reduce the possibility of misinterpretation.

SECTION 320-14. Rigid nonmetallic conduit is now acceptable as protection for open wires that are subject to physical damage because of their location on walls or ceilings.

Article 321
Messenger Supported Wiring

This new Article fills a void that has existed for many years. Cable assemblies consisting of one or more insulated conductors twisted together and supported by a bare messenger are usually referred to as duplex, triplex, or quadruplex. These factory-assembled cables have a long record of satisfactory and reliable service as service drops. Triplex and quadruplex cables have also been used for overhead wiring of parking lot lighting, playground lighting, and for supplying one building from another. In the medium-voltage (above 2001 volts) range, self-supporting aerial cables are used in industrial plants.

SECTION 321-1. Notice that the cables can be supported by a number of different methods. Also, the cable may be assembled in the factory or field assembled. Factory-assembled cable is preferred because it is difficult to uniformly band a long run of single conductors together in the field.

SECTION 321-2. Where messenger-supported cables are installed outdoors, the rules for outside wiring apply. General requirements for all wiring methods that have a bearing on this type of system must also be followed.

SECTION 321-3. Types of cables that can be supported from a messenger are listed in Part (a).

In addition to the cable types mentioned in Part (a), messenger-supported medium-voltage cables can be used in industrial plants where competent individuals maintain the messenger-supported wiring.

Cable and conductor types that are permitted in Class I, II, and III hazardous locations may be supported by a messenger according to Part (c).

SECTION 321-4. The only locations where this cable cannot be installed are hoistways and areas that would subject the cable to severe physical damage.

SECTION 321-5. The ampacity of the conductors in the assembly is determined from Tables used for conductors in raceways, not free air. There are no exceptions to this requirement, even where only two conductors are involved. Also, four-conductor cable (quadruplex) must be derated in accordance with Note 8 to the Tables if all conductors carry current.

SECTION 321-6. Conductors supported by the messenger cannot be in contact with the supports or any other object. Although spacing between end supports is not defined, intermediate support must be provided to eliminate excessive strain on the messenger or tension on the conductors.

SECTION 321-7. The messenger must be grounded under all conditions. The Sections referenced cover grounding of service raceways and other metal enclosures.

SECTION 321-8. Splices and taps are permitted where made with suitable materials and approved methods.

Article 324
Concealed Knob-and-Tube Wiring

Since this wiring method is generally prohibited except for additions to existing installations or by special permission, there is no demand for change.

SECTION 324-8. This Section began with "Where practicable, conductors shall be run singly on separate joists, studs or rafters." Since the phrase "where practicable" creates enforcement problems, the sentence is deleted.

Article 326
Medium Voltage Cable

SECTIONS 326-3 and **326**-4. Previously, type MV cable could not be used as a messenger-supported cable; now it can. The change in these two Sections moves messenger-support cable from a prohibited use to a permitted use.

SECTION 326-6. The Exception is added to point out that the ampacity of medium-voltage cables is modified when such cables are installed in cable trays.

Article 328
Flat Conductor
Cable Type FCC

Proposed requirements for this wiring system were submitted for inclusion in the 1978 NEC, but were not adopted. After publication of the 1978 Edition, a Tentative Interim Amendment was processed in accordance with NFPA procedures, and this proposal was accepted. Since all TIAs must be processed during the next revision cycle of a standard and obtain committee approval, Article 328 is new in this issue of the Code even though this wiring method had prior official acceptance. The description that follows, although general in nature, relates to one manufacturer's product.

The conductors are No. 12 AWG copper flattened to a width of about ½ inch and a thickness of about 0.01 inch. These conductors run parallel to each other and are sealed in two layers of plastic film. For three conductors the width of the ribbon is about 2 ¾ inches and the total thickness is approximately 0.025 inch. The bottom shield is either plastic or metal about 10 mils thick. The top shield is metal, also about 10 mils thick, with a width adequate to completely cover and protect the conductor ribbon. Rapid addition shows that the total thickness of this "sandwich" is about 45 mils, or less than 1/16 inch. Where cables ends are spliced together or a branch is tapped from the main run of cable, splicing devices increase this thickness slightly. The thickness is doubled when the cable is folded over on itself to change direction.

Connectors are provided to make end-to-end splices and 90° taps. But even at splice points the thickness is not increased enough to make a lump that can be seen on the surface of carpet squares.

End caps are used to seal the exposed conductors at the end of each run. Finally, specially designed junction boxes are used to connect this cable to a conventional wiring system, and floor fittings are provided to connect receptacles to the flat conductor cable. Now let's look at some of the specific requirements.

SECTION 328-1. Only carpet squares (nothing else) can be placed over FC cable. This wiring method is designed for installation under carpet squares, and this is the only floor covering material that is acceptable.

SECTION 328-3. Since this is a branch-circuit wiring system, it must conform to the requirements for branch circuits. Load calculations for this wiring system are covered by Article 220. Overcurrent protection, grounding, and general requirements for all wiring methods as covered by Articles 240, 250, and 300 apply to flat conductor cable.

SECTION 238-4. FCC systems are permitted for general-purpose, appliance, and individual branch circuits.

According to Part (b), most types of floors are suitable. Brick, exposed aggregate, or broken tile floors are some types that should be avoided. Rough floor surfaces should be repaired and any holes patched. Grease and oil must be removed and the floor must be cleaned. Surface dust and dirt will result in poor adhesion of the cable to the floor.

Notice that FC cable can be installed on walls, in damp locations, and on heated floors that have a surface temperature exceeding 86°F.

SECTION 328-5. This wiring system is not permitted in wet locations, outdoors, in corrosive areas, or in hazardous locations. It cannot be installed in residences, schools, and hospitals.

SECTION 328-6. Voltage is limited to 120/240 volts single-phase or 208Y/120 volts three-phase. Branch-circuit ratings are 20 amperes maximum on multioutlet circuits and 30 amperes for single-outlet circuits.

SECTION 328-10. Carpet squares placed over the cable system must be attached to the floor with release-type adhesives to permit removal and replacement without damage to the flat cable or carpet squares.

SECTION 328-11. All cable ends and connections must be sealed to prevent the entrance of moisture and fluids. Plastic patches and adhesive are used for sealing.

SECTION 328-12. It is important that the top metal shield completely cover the cable, because this grounded metal cover protects the conductors

Section 328-37   51

from penetrations by tacks, paper clips, pins, and any other metal objects that might drop on the carpet and be walked on.

The bottom shield material is not specified. It is laid on the floor before the cable is installed and helps smooth out imperfections in the floor which could physically damage the cable.

SECTION 328-13. All metal transition enclosures, receptacle housings, and metal shields must be grounded through fittings, studs, or connectors to assure a grounding path whose resistance is not greater than one conductor in the cable. This resistance is difficult to measure, but if all parts of the grounding system are connected as specified by the manufacturer, a low-resistance ground will result.

SECTION 328-14. The grounding conductor in the FC cable must be connected to the grounding shield at all receptacle locations. Interconnection of the equipment-grounding conductor in the cable with the metal shield results in a redundantly grounded shield which further reduces the electrical resistance of the shield.

SECTION 328-15. The transition assembly is a special junction box containing a terminal assembly which makes it possible to connect the flat conductor cable and shield to a conventional wiring system.

SECTION 328-16. Receptacle pedestals must be securely fastened to the floor. The flat conductor cable must also be secured to the surface.

SECTION 328-17. Wherever cables cross each other they must be separated by a layer of grounded metal shielding.

SECTION 328-18. Should the cable thickness exceed 90 mils (about 3/32 inch) because of splices, taps, crossovers, etc., the edges must be tapered. This reduces the probability of stumbling.

SECTION 328-19. Alterations can be made to the system and cables no longer needed can remain energized and in place provided that ends now exposed because of receptacle removal are properly sealed.

SECTION 328-20. The cable has to be laid out so that polarity is maintained at splices, taps, and receptacle terminals.

SECTIONS 328-32 through **328-37.** These are construction requirements of a general nature; they do not specify dimensions or materials.

Notice that the cable must be marked on both sides because either surface can be attached to the floor. Color coding is used to identify the conductors.

White, black, and green surface markings are used to identify conductors in a three-wire cable.

The remaining Sections contain requirements which are similar to those already reviewed.

Article 330
Mineral-Insulated, Metal-Sheathed Cable

There are no substantial changes in this Article.

Article 333
Armored Cable

There are no substantial changes in this Article.

Article 334
Metal-Clad Cable

SECTION 334-11. To permit a smaller bending radius for multiconductor shielded cable, the method for determining minimum bending radius is changed. Now you can use either 12 times the overall diameter of one conductor or seven times the overall diameter of the multiconductor cable, whichever is greater.

SECTION 334-13. The ampacity of conductors in type MC cable rated 2000 volts or less is obtained from Tables 310-16 through 310-19. The type of insulation on the contained conductors is used to select the proper column.

Section 310-15 and the related Tables are used to obtain ampacities for cables rated over 2000 volts. The Exception points out that ampacities taken from the Tables are subject to correction factors where installed in cable trays.

Article 336
Nonmetallic-Sheathed Cable

SECTION 336-3. To determine the number of floors in a structure built on a sloping grade, the first floor is defined as the lowest level designed for human habitation and having at least 50 percent of its perimeter (floor perimeter) level with or above finished grade. Previously, it was not clear whether the determination was made at floor level or by measurement of exposed exterior walls.

SECTION 336-4. The reference to Note 8 of Tables 310-16 through 310-18 is added to emphasize the importance of derating where cables are bunched. Very little concern has been given to that part of Note 8 which states: "Where single conductors or multiconductor cables are stacked or bundled without maintaining spacing and are not installed in raceways, the maximum allowable load current of each conductor shall be reduced as shown in the above table." With the emphasis being placed on thermal insulation, bundled conductors surrounded by this material are sure to overheat if not derated. This potentially dangerous situation can be avoided by planning the wiring layout in advance. For example, at the distribution panel some cables can be routed into adjacent stud spaces instead of having all the cables enter into the same stud space occupied by the panelboard. Where bundling cannot be avoided, derate.

SECTION 336-13. Wiring devices designed for installation without outlet boxes are acceptable provided that the device is secured to a bracket that is fastened to a structural member of the building.

Article 337
Shielded Nonmetallic-Sheathed Cable

There are no substantial changes in this Article.

Article 338
Service-Entrance Cable

There are no substantial changes in this Article.

Article 339
Underground Feeder
and Branch-Circuit Cable

SECTION 339-1. The minimum-size conductors in UF cable is No. 14 copper or No. 12 aluminum or copper-clad aluminum. The maximum size regardless of conductor material is No. 4/0.

Article 340
Power and Control Tray Cable

SECTION 340-4. Type TC (tray cable) can now be used for communications circuits in addition to power, lighting, and signal circuits.

By rewriting this Section, some restrictions on where the cable can be used have been lifted. In the past this wiring method was permitted only where conditions of maintenance and supervision assured that only qualified persons would service the installation. Now qualified individuals are required only where the cable is installed in cable trays in hazardous locations. All other permitted uses are no longer restricted to industrial occupancies where qualified people maintain the cables.

Article 342
Nonmetallic Extensions

There are no significant changes in this Article.

Article 344
Underplaster Extensions

SECTION 344-2. Rigid nonmetallic conduit is added to the list of acceptable materials for underplaster extensions.

Article 345
Intermediate Metal Conduit

SECTION 345-8. The old language in this Section could be interpreted to mean that the only way to remove rough edges from the cut end of conduit was with a reamer. Actually, any tool that does the job is acceptable, and the revision makes this clear. Similar revisions are made for rigid metal conduit and electrical metallic tubing.

Article 346
Rigid Metal Conduit

SECTION 346-1. Part (b) states that the use of dissimilar metals in a conduit system should be avoided to reduce the possibility of galvanic corrosion. Formerly, the Exception permitted aluminum fittings and enclosures with steel conduit. This Exception is now expanded to permit steel fittings and enclosures with aluminum conduit.

Article 347
Rigid Nonmetallic Conduit

SECTION 347-1. Fiberglass epoxy conduit is added to the list of nonmetallic conduits suitable for underground use. This material is superior to some other nonmetallic conduits because its coefficient of thermal expansion is comparable to metal, it has good resistance to chemicals normally found in soils, and it has high tensile strength.

SECTION 347-2. Rigid nonmetallic conduit does not have to be encased in 2 inches of concrete for voltages over 600. In the 1978 NEC this conduit could be used with voltages up to 600, but above this voltage it had to be concrete-encased. The one exception to this statement applies to nonshielded cables. Where these cables are installed underground in rigid nonmetallic conduit the raceway must be encased in 3 inches of concrete.

SECTION 347-9. The language is improved to require expansion joints for rigid nonmetallic conduit. Where this material is subject to temperature variations, expansion joints are necessary to prevent buckling or pulling apart at couplings. An adequate number of expansion joints in a conduit run will eliminate these problems.

Article 348
Electrical Metallic Tubing

SECTION 348-11. Cut ends of electrical metallic tubing can be made smooth by using any tools, including a reamer.

Article 349
Flexible Metallic Tubing

There are no substantial changes in this Article.

Article 350
Flexible Metal Conduit

SECTION 350-3. It is now clear that ⅜-inch flexible metal conduit—not more than 6 feet long—can be used to connect a recessed lighting fixture to branch-circuit wiring.

SECTION 350-6. The title is changed from "bends in concealed work" to "bends." "Concealed" is also removed from the text. Any random bending of flexible metal conduit that is fished in finished walls cannot be seen, so why use the word "concealed"? Besides, the number of bends must be restricted, regardless of whether the conduit will be concealed or remain exposed.

Article 351
Liquidtight Flexible Conduit

The title is changed because this Article now covers metallic and non-metallic types.

SECTION 351-10. This Section deals with the number of bends allowed between boxes and fittings and has undergone the same changes as Section 350-6.

SECTIONS 351-21 through 351-27. All Sections from here to the end of the Article cover a new wiring method, which is called "liquidtight flexible non-metallic conduit." This nonmetallic raceway is recognized for limited use in industrial applications. Although in use for a number of years in the machine tool industry, it did not have NEC status until now.

SECTION 351-22. Liquidtight flexible nonmetallic conduit can be loosely described as a tube within a tube. It is made up with a seamless inner core bonded to an outer cover having reinforcing layers between them.

SECTION 351-23. This conduit can be used where flexibility is necessary, and where conductors must be protected from vapors, liquids, or solids. Liquidtight flexible nonmetallic conduit cannot be used where subject to physical damage; where exposed to high temperatures, either external or internal; where voltage of contained conductors is above 600 volts; or where the length is in excess of 6 feet unless approved for special installations. On many large machine tools the length restriction may not be realistic; therefore, the escape clause "unless approved for special installations" is included.

SECTION 351-24. The maximum size is limited to 1 ½ inches.

SECTION 351-25. The number of conductors permitted in the conduit is obtained from the same Tables that are used to determine raceway fill for conduit and tubing.

SECTION 351-26. Only terminal fittings designed for liquidtight flexible nonmetallic conduit are acceptable.

SECTION 351-27. Where required, an equipment-grounding conductor has to be pulled in with the circuit conductors. There is no provision for an external equipment-grounding conductor.

Article 352
Surface Raceways

There are no substantial changes in this Article.

Article 353
Multioutlet Assembly

There are no substantial changes in this Article.

Article 354
Underfloor Raceways

There are no substantial changes in this Article.

Article 356
Cellular Metal Floor Raceways

There are no substantial changes in this Article.

Article 358
Cellular Concrete Floor Raceways

There are no substantial changes in this Article.

Article 362
Wireways

SECTION 362-5. This Section is rewritten and the language improved for clarity. The first paragraph allows 30 current-carrying conductors at any cross section along the wireway without derating. Additional conductors for control and signaling which do not carry current except for motor starting are not counted as part of the 30 current-carrying conductors. However, the total cross-sectional area of all conductors (current-carrying, control, and signaling) cannot exceed 20 percent of the area of the wireway.

Where there are more than 30 current-carrying conductors along any cross-sectional area of the wireway, Note 8 to Table 310-16 through -19 applies and all contained conductors have to be derated by the percentage corresponding to the total number of current-carrying conductors at any cross section of the wireway. Stated another way, if the maximum number of current-carrying conductors at any cross section of the wireway is 30, no derating is necessary, but if 31 conductors are present, they all must be derated to 60 percent of the values shown in the Amacity Table. Here again, the area available for conductors is limited to 20 percent of the internal cross section of the wireway.

The second Exception permits more than 30 current-carrying conductors without derating where the contained conductors supply border lighting and stage pockets in theaters and similar occupancies. However, the 20 percent fill restriction still applies.

Vertical wireways installed in hoistways for elevators are permitted to be filled to 50 percent of their cross-sectional area by Section 620-32. Exception No. 3 rightfully points this out.

SECTION 362-10. Formerly, only metallic wiring methods could be connected to wireways. Now rigid nonmetallic conduit is also recognized. Where equipment grounding is required, the grounding conductor must be connected to a clean (unpainted) surface in the wireway with a pressure lug, clamp, or other suitable device.

Article 363
Flat Cable Assemblies

This is a branch-circuit wiring system consisting of a flat two-, three-, or four-conductor assembly designed for field installation in metal raceway. Originally limited to supplying phase-to-neutral loads through special tap connection fittings, Section 363-10 is revised to permit phase-to-phase connections.

SECTIONS 363-9 and **363-14.** Formerly, supply connections to, and extensions from, flat cable assemblies had to be made on terminal blocks in specially designed junction boxes. Since individual conductors in the assembly are surface-marked to maintain proper polarity, terminal blocks are not necessary. Connections can be made with screw-on connectors or other conventional splicing devices.

Article 364
Busways

SECTION 364-8. Rigid nonmetallic conduit is now recognized as a wiring method that may extend from a busway. An equipment-grounding conductor must be included in the conduit and terminated according to the requirements of Sections 250-113 and -118.

Flexible cord assemblies can also be used for extensions from busways, and the conductors can be directly connected to the load terminals of the busway plug-in device provided that a suitable strain-relief clamp is used to prevent any pull at the connections.

Article 365
Cablebus

There are no substantial changes in this Article.

Article 366
Electrical Floor Assemblies

There are no substantial changes in this Article.

Article 370
Outlet, Switch and Junction Boxes, and Fittings

SECTION 370-3. The second paragraph, permitting the use of nonmetallic boxes over 100 cubic inches with metallic wiring systems, was a revision that appeared for the first time in the 1978 NEC. To separate this requirement from the basic rule that permits only nonmetallic boxes with nonmetallic wiring systems, the previously existing statement was changed to read "Nonmetallic boxes not over 100 cubic inches shall be permitted only with open wiring on insulators, concealed knob-and-tube wiring, nonmetallic sheathed cable. . . ." As you can see, the revision had the effect of limiting the size of nonmetallic boxes to 100 cubic inches where the wiring method is nonmetallic. This was not the intent. The revised statements now make it clear that any size nonmetallic box can be used with nonmetallic wiring. And nonmetallic boxes over 100 cubic inches with built-in bonding means for all conduit entries are permitted with metal wiring methods.

SECTION 370-6. This Section provides rules for determining the number of conductors in switch, outlet, receptacle, and junction boxes. Part (a)(1) requires reduction of the number of conductors in boxes where fittings and wiring devices are installed. Formerly, one conductor had to be deducted from the number allowed by Table 370-6(a) where fixture studs, cable clamps, or hickeys were contained in the box. There were some people who required a reduction for each type of fitting, one for clamps and one for a fixture stud, or a total of two. Under my interpretation, one conductor was deducted. Actually, the language was not precise and this caused varying interpretations. The new wording makes it clear that one conductor must be deducted for cable clamps. An additional deduction of one conductor must be made for a fixture stud or hickey. To show how this works, let's find out what size box is required for three No. 12, two-conductor with ground NM cables, one No. 12, three-conductor with ground NM cable, and a fixture stud in a ceiling outlet box. Adding one conductor for the clamps, one for the stud, and one for the grounding conductors to the nine insulated conductors results in a total of 12 No. 12s. Now referring to Table 370-6(a), we find that a 4 × 2 1/8 inch square box is adequate, since it can accommodate 13 conductors. To summarize: The number of conductors shown in the table must be reduced by one if the box contains cable clamps, by two if the box contains both cable clamps and a hickey, and by two if the box contains cable clamps and a fixture stud.

The marked volume of extension rings, domed covers, etc., connected to outlet boxes can be added to the volume of the box in determining total cubic-inch capacity. Where this extension ring has the same dimensions as the box to which it is connected, no marking is necessary because the volume of the extension can be found in the Table.

SECTION 370-7. It is not necessary to clamp nonmetallic-sheathed cable to a single gang nonmetallic box where the cable is secured within 8 inches of the box and at least 1/4 inch of the sheath extends into the box. Under all other conditions any wiring methods permitted with nonmetallic boxes must be either externally or internally secured to the box.

SECTION 370-11. Drywall and plasterboard are added to plaster as the types of building surfaces that must be repaired where broken or incomplete. The former reference was just to plaster but should have included similar materials.

SECTION 370-13. Where nails are used to support a box, they must not be more than ¼ inch from the edge or back of the box. The change here allows ¼ inch from the edge. Better support of 3½-inch-deep boxes is realized when the nails are located near the edges instead of near the back.

SECTION 370-17. Floor boxes must be specifically listed for this application. The Exception is rewritten to conform to the standard format. Under this Exception, general-purpose flush receptacle boxes are acceptable.

SECTION 370-18. Covers for pullboxes, junction boxes, and fittings must be compatible with the box or fitting construction and suitable for the conditions of use. This detailed explanation is better than the old terminology "approved for purpose."

SECTION 370-20. Part (d) is new and requires a grounding means for an equipment-grounding conductor in each metal box designed for use with nonmetallic raceways or nonmetallic cable wiring. A tapped hole, stud, or similar arrangement that secures an equipment-grounding conductor to the box is satisfactory.

Article 373
Cabinets and Cutout Boxes

SECTION 373-6. There are major changes in wire bending spaces for terminal connections. Although this Article deals with cabinets and cutout boxes, these rules are also applicable to wire bending space in switches and panelboards.

TABLE 373-6(b). Table 373-6(b) is new and applies where the conductor leaves through the wall opposite the terminal. For a 200-ampere enclosed safety switch with No. 3/0 lugs on the top and bottom, the distance from the end of the lug to the top of the enclosure must be 6 ½ inches. This same space must be provided for the bottom lugs.

A sampling of the former minimum bending space with the dimensions now required by Table 373-6(b) shows some radical changes. For 500 MCM conductors the bending space is increased from 6 inches to 14 inches; for 1250 MCM the increase is from 10 inches to 22 inches. In the smaller sizes the increases are not as dramatic, but will result in a reduction of skinned knuckles, cuts, and bruises. With more bending space the probability of damaged insulation is also reduced.

A proposal to increase the wire bending spaces at terminals where conductors had to be bent or deflected more than once was submitted as a revision for the 1978 Edition of the NEC. The proposal was held on the docket and a committee was appointed to study the problems associated with inadequate bending space in enclosures containing terminals. Insufficient bending space can result in conductor insulation damage, poor connections due to shifting of the conductor strands and misalignment of the lay, and excessive strain on the terminals. Hand injuries such as cuts, nicks, and skin abrasions have also occurred during the battle to get the conductor into the terminal.

With the help and guidance of a steering committee, Underwriters Laboratories, Inc., undertook a fact-finding study to obtain factual data on wire bending space needed to make wire connections at terminals. The results indicated that wire bending spaces listed in Table 373-6(a) were not sufficient for offset or S bends. Difficulty was also encountered in terminating conductors that go straight into the terminals. Where a conductor enters the enclosure directly opposite the terminal, the wire must be bent in a semicircle so that it can be inserted in the terminal barrel, an exercise requiring extreme strength on the part of the electrician where large conductors are involved. With this background information, let's take a close look at the changes.

Part (b) says that either Table 373-6(a) or -6(b) is applicable for determining wire bending space at terminals. Table -6(a) is used where the conductor does not enter or leave the enclosure through a wall opposite its terminal. An example of this is where the conductors enter through a side wall and terminate vertically on lugs. In this instance the wire is bent about 90 degrees from horizontal to vertical. Since only one bend occurs in the wiring space around the terminal, the dimensions in Table -6(a) apply. The Exception permits the bending space opposite the terminals to be as shown in Table -6(a), where the conductors enter or leave the enclosure in the opposite wall provided that the point of entrance or exit of the conductors is into a wiring space whose dimension conforms with Table 373-6(b). Where conductors are installed in this manner a double bend is necessary, but additional bending space is provided at the conductor entry or exit point.

Paragraph (2) of Part (b) simply states that Table 373-6(b) applies where the conductor enters or leaves the enclosure through the wall opposite its terminal: a general-use (safety) switch is a good example of this type of construction.

Remember, in the beginning of this discussion I stated that the changes were a radical departure from existing standard practices. Redesign and retooling of enclosures for switches and panelboards are necessary. Therefore, it will be some time before products with the enlarged bending spaces appear in electrical distributors' inventories.

Article 374
Auxiliary Gutters

There are no substantial changes in this Article.

Article 380
Switches

SECTION 380-6. The language is improved to make it clear that a locking device is necessary which prevents closing of the blades of a knife switch designed for inverted installation. Similar changes are made for double-throw switches that are mounted in a vertical position. With this type of switch a locking device is required to hold the blades in the open position.

SECTION 380-14. Snap switches rated 20 amperes or less and directly connected to No. 10 or No. 12 aluminum alloy conductors must be marked CO/ALR. This marking usually appears on the mounting yoke and is visible when the plate is removed. Part (c) is new, but this requirement appeared under the subheading AC General-Use Snap Switches. The revision simply means that CO/ALR switches are not necessarily ac general-use switches; they may also be designed as ac-dc general-use snap switches.

SECTION 380-18. This is a new Section which requires wire bending space according to Table 373-6(b). As mentioned previously, these increased bending spaces were found to be necessary as a result of a fact-finding investigation conducted by UL.

Article 384
Switchboards and Panelboards

SECTION 384-2. This Section, which formerly contained references to other Articles, is moved to Part (b) in Section 384-1.

Switchboards, panelboards, and distribution boards must be located in rooms or spaces that are reserved for the exclusive use of electrical equipment. At first glance the wording seems to be very restrictive and will cause many varying interpretations. If this electrical equipment is not located in a room by itself, the space occupied by the equipment must be free of any foreign piping, ducts, etc., which through any failure can damage or interfere with the safe operation of the electrical equipment. Water piping (hot and cold), drains, sanitary pipes and cleanouts, chilled water pipes, and steam pipes are examples of piping systems that cannot pass above the space occupied by switchboards and panelboards. Air-conditioning supply and exhaust ducts are some of the types of duct systems that also cannot be located in the space reserved for electrical equipment covered by this Article.

Since no dimensions are assigned to the space around this type of electrical equipment where it is not located in a room designed exclusively for the purpose, you may ask: How much space is required? This is where varying interpretations will occur, but here are some guidelines. The intent is to protect the electrical equipment from damage caused by leaking pipes, valves, and condensation. Where the contained liquid is under pressure, a leak may result in spraying of the equipment even though the leaking pipe is some distance away. Therefore, some drain and sanitary pipes can be located nearer a switchboard or panelboard than, for example, a steam pipe. Condensation from air-conditioning ducts drops directly below the duct, so ducts and registers cannot be located directly above the electrical equipment. They only have to be far enough away to prevent this occurrence. These examples are cited to point out that there are no simple answers, and judgment must be used in applying this rule.

The first Exception permits control equipment for motorized valves, pumps, etc., to be located adjacent to ducts, pipes, and other equipment foreign to the electrical installation.

Ventilating and air-conditioning equipment which is used for the electrical equipment is permitted in the same space or room by Exception No. 2.

Many industrial plants contain large amounts of piping, with switchboards and panelboards scattered throughout the plant. Where it is impractical to isolate the switchgear from foreign piping, special designs of enclosures are employed to protect the electrical equipment from leaking or faulty piping. This is the reason for Exception No. 3.

The last Exception provides relief for outdoor electrical equipment in weatherproof enclosures. In many chemical plants, outdoor unit substations are scattered throughout the plant site. Although good design dictates locating the electrical distribution equipment in protected areas, large amounts of process piping often prevent the electrical engineer from selecting an ideal location that is free of piping and not subject to physical damage. Exception No. 4 recognizes this problem and permits other safeguards where proper clearances cannot be obtained.

SECTION 384-3. Conductors entering or leaving a switchboard must do so in the compartment in which they terminate. Or stated another way, conductors that terminate in a vertical section of a switchboard must be located in that section. This means that a conduit cannot terminate anywhere on the top, bottom, or side of a switchboard with the conductors running horizontally through the switchboard to an overcurrent device that is located in another section.

This is a good rule, but some hardships have resulted where it was necessary to redesign a switchboard because of unforeseen obstructions, tenant changes, etc., during early stages of construction which had proceeded to the point where conduits were already in place and floor slabs poured. Usually, conduits can be stubbed out in proper relation to the switchboard by following the manufacturer's shop drawings, but when a change is made that requires redesign of the switchboard, some conduits are usually in the wrong place. To prevent having to break up concrete and reposition conduits, an Exception is added to allow conductors to travel horizontally through the board provided that the conductors are isolated from the buses by a barrier. Some switchboard manufacturers can provide additional wiring space at the top of the board. This space is usually separated from the bus sections by a metal bottom. Since the Exception does not mention metal, the barrier can be of nonmetallic material.

According to Part (f), the phase conductor having the higher voltage to ground on a three-phase four-wire delta system must be the B phase and occupy the center position on switchboards and panelboards. The grounded conductor (corner-ground) on a three-phase three-wire delta system is also referred to as the B phase. To separate the two, the three-phase, four-wire delta system is mentioned.

SECTION 384-9. This Section, dealing with conductor insulation in switchboards, is rewritten to spell out the requirements without listing all the acceptable types of insulation, as was previously done.

SECTION 384-16. Part (a) requires overcurrent protection for lighting and appliance panelboards to prevent overloading. All of these panelboards must be protected by one or not more than two circuit breakers or two sets of fuses not exceeding the ampere rating of the panel. Exception No. 2 allowed up to six circuit breakers or six sets of fuses as the overcurrent protection for lighting and appliance panelboards where used as residential service equipment. This provision no longer exists for new construction. However, existing panelboards may remain, and any spares can be used to supply new appliances, building additions, or other added loads. With this change split-bus lighting and appliance panels are limited to two mains when installed as service equipment. If a lighting and appliance panelboard with six mains in the upper section is used, it must be protected on the line side by overcurrent devices not exceeding the ampere rating of the panel.

SECTION 384-25. This new Section requires wire bending space at the top and bottom of panelboards based on dimensions in new Table 373-6(b). On top- or bottom-feed panelboards, feeder conductors must have double (offset) bends to connect to the terminals on MLO panelboards. Testing showed that additional space is necessary for proper terminations. Gutter space on the sides of the panels cannot be less than shown in Table 373-6(a) and is based on the largest conductor terminated in that space.

The first Exception allows either the top or bottom bending space to be smaller for a lighting and appliance branch-circuit panelboard with a bus capacity of 225 amperes or less. The space for terminating the supply conductors is determined from Table 373-6(b) and the opposite space from Table 373-6(a). For a 225-ampere panel with 300 MCM aluminum lugs on the top, the distance from the edge of the lugs to the top wall has to be 10 inches, while the bottom space only has to be 5 inches.

The second Exception recognizes another type of construction. Either the top or bottom wire bending space can be reduced to the values listed in Table 373-6(a), where one side gutter space is sized according to Table 373-6(b) for the largest conductor terminated in that space. To see how this works let's assume a 400-ampere panel with main lugs at the top and a 100-ampere three-pole circuit breaker as the largest overcurrent device in the panelboard. If the main lugs are doublebarrel for No. 3/0 copper conductors, the distance from the lugs to the top wall must be 6 ½ inches. If the terminals on the 100-ampere circuit breaker will accept a No. 1/0 conductor, the distance to that side wall must be 5 ½ inches.

SECTION 384-27. An equipment-grounding bar is required in all panelboards wherever equipment-grounding conductors are used. This terminal bar cannot be connected to the neutral bus except where the panelboard is part of the service equipment. Since Section 250-24 requires grounding the neutral in a separate building—under most conditions—the neutral bar and equipment-grounding terminal bar can be interconnected. This is the sense of new Exception No. 2.

FOUR
Equipment for General Use

Article 400
Flexible Cords and Cables

SECTION 400-4. Usually, with every Code edition there is an introduction of new insulating compounds for flexible cords. This edition is no exception. Flexible cords with insulation and outer jackets previously referred to as rubber are now designated "thermoset," because cords classified as "rubber" employing elastometric (rubber) compounds which are not thermoset are not listed by UL.

Flexible cords having conductor insulation and an outer jacket of thermoplastic elastomers are listed as SPE types under the heading "All Elastomer (Thermoplastic) Parallel Cord." Type SJE is a Junior Hard Service Cord made of thermoplastic elastomer. Then there are some new Hard Service Cords made of Thermoset Compounds that have oil-resistant insulation in addition to an oil-resistant jacket. Types SOO and STOO are examples. Note No. 6, covering construction of elevator traveling cables, is revised to permit nonmetallic fillers as necessary to maintain concentricity. A steel supporting cable is not necessary where the distance between supports is less than 100 feet. Materials other than steel can be used for support where the traveling cable is subject to excessive moisture or corrosive vapors or gases.

The language in the second paragraph is changed to allow more than one pair of telephone wires and more than one shielded coaxial cable. Since communication facilities and monitoring within an elevator car are increasing, more than one communication circuit and video circuit are often necessary.

SECTION 400-5. The conductor metal suitable for use in flexible cords and cables is copper. This means that stranded aluminum, copper-clad aluminum, or any other metals are not suitable as current-carrying conductors.

Where multiconductor flexible cords are used, the ampacity of the conductors must be derated according to the derating factors shown.

Formerly, the ampacities of conductors in a flexible cord or cable had to be reduced to 80 percent of the values shown in Table 400-5 where more than three conductors carried current, and the derating factor remained at 80 percent even though the number of current-carrying conductors increased to nine, for example. In other words, the derating factor did not change regardless of the number of current-carrying conductors in the multiconductor cable. With the addition of the Derating Factor Table, the allowable load current decreases as the number of current-carrying conductors increases. Now a flexible cord or cable containing nine current-carrying conductors must be derated so that the load current on each conductor does not exceed 70 percent of the value shown in Table 400-5.

In Table 400-5 the ampacity of No. 20 wire is added for elevator cables.

SECTION 400-9. A parenthetical phrase is added to make it clear that only hard service cords such as type SO, SE, ST, STO, etc., are acceptable for splicing. All others, and hard service cords smaller than No. 12, cannot be spliced.

SECTION 400-14. Protection from damage is required where flexible cords pass through sheet metal holes in covers and enclosures.

SECTION 400-31(a). Conductors used in portable cables over 600 volts must be No. 8 AWG copper or larger. Notice that the conductor material specified is copper.

Article 402
Fixture Wires

TABLE 402-3. Wire sizes for heat-resistant fixture wire are now correctly stated for sizes Nos. 18 through 14. Multiconductor type KF-2 and KFF-2 fixture wires are acceptable in Class 1, 2, and 3 circuits and fire protection signaling systems. Types HF and HFF are also recognized for use in Class 1, 2, and 3 systems.

SECTION 402-11. Since fixture wires are used as branch circuit conductors for Class 1 systems and fire protection circuits, the Exception states this act.

Article 410
Lighting Fixtures, Lampholders, Lamps, Receptacles, and Rosettes

SECTION 410-4. Lighting fixtures installed in nonresidential cooking hoods must be specifically marked and identified for this application. A number of manufacturers have fixtures that are suitable for installation in commercial cooking hoods.

Part (d) permits chandeliers, hanging fixtures, and swag lamps over bathtubs in bathrooms where the lowest part of the fixtures is not less than 8 feet from the top of the tub. Whether a fixture installed as indicated has to be "suitable for damp locations" depends on many variables, such as height of the fixture, size of the bathroom, ventilation, etc. Having the fixture at least 8 feet above the rim of the bathtub assures that relamping will not be attempted by a person standing on the edge of the tub. This mounting height also eliminates chandeliers from most bathrooms because the ceiling must be approximately 12 feet above the floor.

Although not required, additional safety from electrical shock can be obtained by connecting the lighting fixture to a branch circuit protected by a ground-fault circuit-interrupter.

SECTION 410-8. If a lighting fixture is installed in a clothes closet, 18 inches of clear space from combustible materials that are stored in the closet must be maintained. Where this clear space is not available because of the size or shape of the closet, a recessed fixture with a solid lens could be installed. Now a fluorescent fixture is also acceptable provided that a 6-inch clear space separates the fixture from combustible materials stored in the closet. A single-tube fluorescent fixture mounted on the ceiling above the door of a 24-inch-deep closet satisfies the 6-inch spacing requirement and provides better lighting inside the closet than does a recessed fixture.

SECTION 410-14. That part of this Section which formerly permitted cord-and-plug-connected fluorescent and mogul-base electric-discharge lighting fixtures is moved to Part (c) of Section 410-30. Other editorial rearrangements of requirements are made in this Article to place rules on the same subject in the same Sections.

SECTION 410-16. Part (b) is relocated from Section 410-30(a).

Lighting fixtures installed in suspended ceilings must be secured to ceiling framing members. It is now clear that the fixtures must be fastened to the suspension members by mechanical means, which include bolts, screws, rivets, and clips. A great deal of difficulty may be encountered in trying to find out which clips are suitable for the fixture, T-bar combination.

Parts (c) and (e) are moved from old Sections 410-40 and 41, respectively.
New Part (g) is relocated from former Section 410-14.

SECTION 410-20. Lighting fixtures with exposed metal parts must be provided with a stud, lug, or pigtail for connecting an equipment grounding conductor. Because some redesign and retooling are required, this rule does not become effective until April 1, 1982. This is a welcome change that will eliminate makeshift grounding of metal lighting fixtures supplied by nonmetallic wiring methods. It is also difficult to properly ground fixtures mounted on plastic boxes. This revision provides a uniform and improved means for grounding lighting fixtures.

SECTION 410-24. Instead of saying that conductor insulation must be suitable for damp, wet, or corrosive locations, terms that are not all-inclusive, the language is changed to state that the fixture wire insulation shall be suitable for the environmental conditions.

SECTION 410-28. Parts (c) and (d) are moved from old Section 410-30(b) and (c).

Parts (e) and (f) are relocated from former Section 410-26(a) and (b).

SECTION 410-29. Where showcases are cord-connected, the secondary conductors for electric-discharge lighting must not leave the showcase in which they originate. (Relocated from Section 410-72.)

SECTION 410-42. This Section requires polarized or grounding-type attachment plugs on all floor, desk, table, and other types of portable lamps. An Exception appeared in the 1978 Edition of the NEC, which stated: "Nonpolarized attachment plugs shall be permitted until January 1, 1980." Since this date for enforcement has passed, the Exception is removed and the rule is now in effect.

SECTION 410-56. A sentence is added in Part (c) to require complete coverage of the opening around receptacles. As mentioned previously, Section 370-11 requires repairs to plaster and drywall where broken around outlet boxes.

Part (e), covering receptacles with a feature that prevents insertion of a nongrounding attachment cap, was accepted as a Tentative Interim Amendment to the 1978 NEC.

Fifteen- and 20-ampere receptacles directly connected to No. 10 or 12 aluminum alloy conductors must be marked CO/ALR.

SECTION 410-57. Standpipes for floor receptacles to allow operation of floor-cleaning equipment without damaging the receptacle was formerly a requirement for cord-connected showcases.

SECTION 410-58. We now have a rule in Part (e) which prohibits installing a grounding-type attachment cap on a cord that does not have a grounding conductor. Connecting a three-wire grounding-type cap to a two-wire cord is deceiving and could give the user of the tool or appliance a false sense of security.

SECTION 410-65(c). Effective April 1, 1982, some recessed incandescent lighting fixtures must have built-in thermal protection to prevent excessive heating.

The energy shortage and high cost of heating fuels have increased the use and thickness of thermal insulation in ceilings. Where insulation is added after recessed incandescent fixtures are installed, dangerous overheating occurs and a potential fire hazard exists because the fixtures are now covered with thermal insulation.

A few years ago Underwriters Laboratories, Inc., began testing surface-mounted incandescent fixtures for installation on insulated ceilings. Fixtures passing this test bear the marking IC, meaning Insulated Ceiling. UL is also testing and labeling recessed incandescent fixtures that can be covered by thermal insulation with no ill effects. These fixtures are identified as suitable for this application. As with all recessed incandescent lighting fixtures, maximum lamp wattage and lamp type (where necessary) are marked in the fixture so that the marking is visible during relamping. Replacement with higher-wattage lamps or a type of lamp other than specified is common practice and can result in fires. My advice is to obey the lamp ratings on the fixtures.

Recessed incandescent fixtures designed for and installed in concrete do not need thermal protection according to Exception No. 1.

Specially marked recessed incandescent fixtures that do not overheat when completely covered by thermal insulation are also exempt from the built-in thermal protection requirement by the second Exception.

SECTION 410-66. This is the Section that requires a clearance of 3 inches between a recessed fixture and thermal insulation. An added phrase recognizes recessed fixtures that have been tested and found suitable for installation within thermal insulation.

SECTION 410-73. A requirement for integral thermal protection of ballasts in indoor recessed high-intensity discharge lighting fixtures appears in new Part (f). Here again the concern is added thermal insulation after the fixtures are in place. An enforcement date of January 1, 1982, is added to allow for redesign and testing.

Article 422
Appliances

SECTION 422-8. Built-in dishwashers, trash compactors, and kitchen waste disposers are permitted to be cord-and-plug-connected under defined conditions. Type SJ cord is added to the list of acceptable types.

SECTION 422-14. Storage-type water heaters with a capacity of not more than 120 gallons were considered as continuous loads in the previous edition of the Code. Since a continuous load is defined as one that is expected to remain energized for 3 hours or more, it was not proper to assign this term to water heaters. The length of time that water heater elements are energized depends on a number of variables, such as storage capacity, number of occupants, living habits, recovery rate, and inlet water temperature, to name a few. According to reports, problems in the field were caused primarily by overvoltage, which resulted in nuisance operation of the branch-circuit overcurrent devices.

Water heater elements are usually rated 4500 watts at 230 volts. Under this set of conditions full-load current is 19.6 amperes. Twenty-ampere overcurrent devices should not open at this value of current, but suppose that the line voltage is 240—what happens then? The full-load current would then be over 20 amperes and the 20-ampere branch-circuit protective device would eventually open. Since voltages above 230 are not uncommon, a revision was made to require increased overcurrent protection and branch-circuit wire size by classifying water heaters as continuous loads. According to Section 210-22(c), a continuous load cannot exceed 80 percent of the branch-circuit rating. From that time to the present, No. 10 wire and 25- or 30-ampere overcurrent devices are required for 4500-watt water heaters. Although this classification of a storage-tank water heater eliminated nuisance opening of

Section 424-22

overcurrent devices, calling this appliance a "continuous load" was not technically correct and arguments arose. These disagreements were compounded when the inspector told the electrical contractor that the ampacity of feeder and service conductors had to be calculated by using 125 percent of the water heater load. This requirement appears in Part (b) of Section 220-10.

To overcome all these problems, Part (b) is rewritten to state simply that the branch-circuit rating shall not be less than 125 percent of the nameplate rating of the water heater.

SECTION 422-16. Instead of listing appliances and the Sections in Article 250 that apply to grounding, the Section is rewritten to refer all appliance grounding to Article 250. A Fine Print Note is added to reference specific appliances to the proper Sections.

Article 424
Fixed Electric Space
Heating Equipment

SECTION 424-3. Part (b) requires branch-circuit conductor ampacity and the rating or setting of overcurrent devices to be not less than 125 percent of the total heater load. When the total load current is multiplied by 1.25, the result is never exactly equal to the ampacity of any of the conductors listed in Table 310-16, or to any of the standard ampere ratings of fuses or circuit breakers listed in Section 240-6. A reference to Section 240-3, Exception No. 1, permits the next-higher standard ampere rating for the branch-circuit overcurrent protection.

SECTION 424-22. Resistance-type heating elements in space heating equipment must be subdivided into loads not exceeding 48 amperes and protected by overcurrent devices not exceeding 60 amperes. The added sentence at the end of Part (b) modifies the preceding statement for heating elements drawing less current than 48 amperes by limiting the overcurrent protection to the next-larger standard ampere rating of a fuse or circuit breaker.

Many electrical inspectors have seen 230-volt single-phase 15-kW furnaces with two 60-ampere double-pole circuit breakers. One circuit breaker protects 10 kW of heating elements and the other protects a 5-kW element and a ¼-hp blower motor. The total load on this second circuit breaker is approximately 25 amperes. Therefore, the ampere rating of the circuit breaker should be 35, not 60. Since there was nothing in Section 424-22 to prohibit this practice, 60-ampere overcurrent devices were used. When questioned, one manufacturer's representative stated that 60-ampere circuit breakers were used in all furnaces, regardless of the kilowatt rating, to expedite production. The change in Part (b) stops this from happening and improves protection of the equipment.

In Part (e) a similiar restriction is placed on the ampere ratings of supplementary overcurrent devices where field-wiring connects the heating elements to the supplementary fuses or circuit breakers.

SECTION 424-34. The Exception permits nonheating leads shorter than 7 feet for heating panels.

SECTION 424-44. The nonheating leads for panels and heating cables embedded in concrete or poured masonry floors can now be protected by PVC conduit where they emerge from the floor. By this time you should realize that the 1981 Edition of the NEC greatly expands permitted uses for nonmetallic conduit.

SECTION 424-72. Resistance-type immersion heating elements contained in an ASME rated and stamped vessel are permitted to be subdivided into 120-ampere loads protected by 150-ampere overcurrent devices. If the subdivided load is less than 120 amperes, the rating of the overcurrent device is determined in the same manner as for the 15-kW electric furnace previously discussed.

A similar change appears in Part (b). The wording is almost identical except that the boiler is not an ASME rated and stamped vessel and the heating elements do not have to be immersion type. Because of these differences, the subdivided loads are reduced to 48 amperes protected by 60-ampere overcurrent devices.

Field-installed wiring between the supplementary overcurrent devices and heater elements must be sized for 125 percent of the load served. Where this calculated current does not equal the ampere rating of a standard-size fuse or circuit breaker, the next larger size is allowed.

SECTION 424-85. Grounding requirements for electrode-type boilers are expanded to cover two different designs. In one type the interior of the boiler is coated with an insulating material so that fault currents cannot pass through the pressure vessel. With this design, the pressure vessel, all exposed metal surfaces, and supply and return piping are grounded. In other designs, where fault current can flow through the pressure vessel, it has to be isolated (see definition) and electrically insulated from ground.

Article 426
Fixed Outdoor Electric De-icing and Snow-Melting Equipment

SECTION 426-1. This Article is rewritten and expanded to include resistance heating, impedance heating, and skin-effect heating.

SECTION 426-2. Most heating systems covered by this Article are engineered and designed as a unit. All components necessary for a complete installation are usually furnished by the manufacturer. Items listed under "Heating System" should all be obtained from the manufacturer of the heating elements or the manufacturer responsible for the design of the system.

Where resistance heating is used, the elements may be tubular (similar to MI cable construction), insulated heating wire, strip heating tape, heating cable, or other similar assemblies.

Some impedance heating systems use steel rods or pipes as the heating elements. For floor heating or outdoor snow melting, the rods are laid in a geometrical pattern to form a mat. A low voltage with relatively high current is supplied to the rods through a two-winding step-down transformer.

Skin-effect current tracing (SECT) has been successfully used in Japan for snow melting on road surfaces and electrical transmission towers. The heating element is a ferromagnetic raceway that contains a single conductor. The insulated conductor runs through the conduit and terminates on the raceway at the far end. One power supply conductor connects to the insulated conductor and the other connects to the raceway. All individual lengths of conduit that form the heat tube are welded.

SECTION 426-3. This is a typical Section that appears in many Articles of the code which deal with electrical equipment. They all generally state that all other applicable rules apply except as modified by any special requirements peculiar to the equipment covered.

SECTION 426-4. The ampacity of branch-circuit conductors supplying fixed outdoor electric de-icing and snow-melting heating systems cannot be less than 125 percent of the full-load current of the heaters. Where a standard-size overcurrent device does not equal the current so determined, the next larger size is permitted.

SECTION 426-10. To assure long life, heating elements and related parts must be suitable for the environment in which they are located. Installation instructions supplied by the manufacturer must be closely followed to assure a safe, trouble-free system that performs as intended.

SECTION 426-11.   Advance planning is often necessary so that all parts of the heating system are located where not subject to physical damage.

SECTION 426-12.   Any exposed surfaces of the heating equipment that attain temperatures above 140°F must be guarded or isolated to prevent accidental contact by individuals.

SECTION 426-13.   Signs must be posted in conspicuous places to indicate that electric heating is present.

SECTION 426-14.   Types of heating systems not covered by these requirements cannot be used unless special permission is granted.

SECTION 426-20.   This Section covers resistance-type heating elements embedded in asphalt, concrete, or masonry. The rules are generally the same as those appearing in previous editions of the Code.

SECTION 426-21.   Exposed de-icing and snow-melting cables must be secured to the surface being heated. Some heating cables will overheat and burn out if not in contact with a surface that acts as a heat sink. However, heating cables that do not overheat when not in contact with the surface being heated are available and should be used for snow-melting and de-icing of roofs, gutters, and downspouts.

Expansion joints should not be crossed with heating cable. Where this cannot be avoided, provision must be made to allow for movement of the cable. This same precaution applies to heating cables installed on flexible structures.

SECTION 426-22.   There are no changes in the requirements for the installation of nonheating leads for embedded heating systems.

SECTION 426-23.   Power supply conductors—nonheating leads—are usually connected to the heating cable or panels by the manufacturer. These leads can be shortened to suit job conditions if identifying markings remain. The leads must be installed in a raceway and terminate in a junction box, where connections to the power supply are made.

SECTION 426-24.   Where splices must be made between heating elements, connectors designed and approved for this application must be used. Connections of nonheating leads to supply conductors are made in a junc-

Section 426-40

tion box suitable for the conditions with splicing devices that are compatible with the conductor metal.

SECTION 426-25. Factory-assembled nonheating leads must be marked at both ends. The marking must identify the manufacturer and include the catalog number plus the electrical rating.

SECTION 426-26. All parts necessary for a complete heating system installation must have corrosion protection that is resistant to expected environmental conditions.

SECTION 426-27. Generally, all exposed noncurrent-carrying metal parts of the complete heating assembly have to be grounded. A metal braid or sheath that covers the heated part of the cable, panel, or unit is required to be grounded by Part (b). After all parts are bonded together with an equipment-grounding conductor, they must be grounded to the distribution panel grounding bar through an equipment grounding conductor sized according to Table 250-95. Note that the equipment grounding conductor is required regardless of the wiring method.

SECTION 426-30. As mentioned earlier, impedance heating results when low-voltage alternating current is applied to a circuit whose conductors are pipes or rods. Any exposed elements must be guarded, isolated, or otherwise insulated to prevent accidental contact by anyone in the vicinity.

SECTION 426-31. The voltage for impedance heating is limited to 30 volts ac, but may be increased to 80 volts where ground-fault circuit-interrupter equipment is installed.

SECTION 426-32. Only dual-winding transformers with a grounded metallic shield between the primary and secondary are acceptable power sources. This assures that primary voltage will not appear on the heating elements.

SECTION 426-33. All current-carrying components must be installed so that there will not be any heating of surrounding metal by induction.

SECTION 426-34. Grounding of the heating system is not necessary where the operating voltage is 30 or less. Above this voltage grounding is mandatory.

SECTION 426-40. The ampacities of conductors as listed in Tables 310-16 and 310-18 are based on three conductors in a raceway in an ambient of 30°C. Since we are dealing with a single conductor in a ferromagnetic conduit, the current-carrying capacity of the contained conductor could be higher than shown in the Tables provided that its temperature rating is not exceeded.

SECTION 426-41. Where necessary to provide pull boxes to facilitate conductor installation, the boxes cannot be buried and covered with earth, asphalt, concrete, etc. The boxes must also be suitable for the environmental conditions.

SECTION 426-42. Section 300-20 prohibits a single conductor in a metal enclosure possessing magnetic properties. Therefore, this Section is necessary to prevent a conflict.

SECTION 426-43. Corrosion protection suitable for prevailing conditions is necessary for all components that make up the skin-effect heating system.

SECTION 426-44. The raceway containing the single conductor has to be grounded at each end. Grounding requirements for a separately derived system as outlined in Section 250-26 do not apply.

SECTION 426-50. For control and protection of the de-icing or snow-melting equipment, a disconnecting means readily accessible to the user is necessary.

SECTION 426-51. A temperature controller with an "off" position must open all ungrounded conductors when in the "off" position. If a positive lockout is provided on the thermostat, it is acceptable as a disconnecting means.

Thermostats without an "off" position do not have to open all ungrounded conductors, but cannot serve as a disconnecting means.

Remote-control thermostats are not acceptable as a disconnecting means.

Line-voltage temperature controllers that include an "on-off" switch are recognized as a disconnecting means where the temperature control contacts do not bypass the switch; the switch opens all ungrounded conductors when moved to the "off" position; and a positive lockout holds the switch in the "off" position.

Article 427
Fixed Electric Heating Equipment for Pipelines and Vessels

SECTION 427-1. These heating systems are usually associated with industrial, chemical processing, and petrochemical plants.

Previously, only resistance heating was covered. The Article is expanded to include impedance heating, induction heating, and skin-effect current tracing.

SECTION 427-2. A pipeline includes all parts of a closed system used to move fluids; a vessel includes all types of containers used for holding fluids.

Notice that thermal insulation, a moisture barrier, and other nonelectrical components may be required for a properly designed and installed heating system.

The Fine Print Note lists the types of resistance heating elements that are available. Exterior heating elements must be in contact with the heated surface to assure proper heat transfer.

The impedance of the pipeline or vessel is the heating element in this type of heating. Alternating current flows in the vessel or pipeline, causing heating of the metal.

Vessels and pipelines are heated by induction by placing a coil around the object to be heated and applying an ac voltage to the coil.

Skin-effect current tracing (SECT) is accomplished by welding a ferromagnetic conduit to a pipeline, installing a single conductor in the conduit, and connecting it to the far end of the conduit. The ac supply is then connected to the free end of the conductor and the conduit. The skin effect of alternating current causes internal heating of the interior surface of the ferromagnetic tube.

SECTION 427-3. All other appropriate rules in the Code apply unless modified by requirements in this Article. Cord-and-plug-connected heaters are recognized but must meet applicable requirements of Article 422—Appliances.

SECTION 427-4. As with other heating systems, the ampacity of branch-circuit conductors and the rating or setting of the overcurrent devices cannot be less than 125 percent of the total heater load. Where the standard ratings of fuses or circuit breakers do not correspond to the calculated current, the next larger standard size is allowed. Regardless of the heater load current, there is no requirement for subdividing into smaller increments as is necessary for fixed electric space heating.

SECTION 427-10. All parts of the heating system must be suitable for all environmental conditions encountered. Since many of these heating systems are specially designed, it is necessary to follow closely the manufacturer's drawings and installation instructions.

SECTION 427-11. Protection from physical damage is essential to assure reliability and reduce accidents.

If the skin-effect current-tracing heating system is supplied from the secondary of a transformer, a separately derived system is created but grounding according to Section 250-26 is not required. Part (d) of this Section permits grounding methods that are prescribed in other portions of the code. In other words, the Fine Print Note refers us to Section 250-26(d), which sends us back to Section 427-48. Catch-22!

SECTION 427-55. The disconnecting means for pipeline and vessel heating systems must be readily accessible to the user. Additionally, the disconnecting means shall be provided with a positive lockout in the off position. There is a difference here compared to the disconnecting means requirements for snow-melting and de-icing heaters in that a lockout feature is not necessary for snow-melting and de-icing.

SECTION 427-56. This Section details various constructions of temperature controllers that are or are not acceptable as disconnecting means for pipeline and vessel heating. Since they are identical to the rules in Section 426-51, no more explanation is necessary.

SECTION 427-57. The rating of overcurrent protective devices is based on the ampacity of the branch-circuit conductors supplying the heating equipment.

Article 430
Motors, Motor Circuits, and Controllers

DIAGRAM 430-1. Editorial revisions of the 1975 NEC clarified the terms overcurrent, overload, and short-circuit, and ground-fault protection. "Overcurrent" in this Article and Article 440 means current resulting from overload, short-circuit, and ground-fault. "Overload" means excessive motor current due to overloading and failure to start. "Short-circuit and ground-fault protection" refers to the motor-branch-circuit protective devices. These terms appear in the Diagram. Additional revisions are made in these Articles by substituting "motor overload" for "motor running overcurrent" and "motor running overloads." Also, "short-circuit and ground-fault" replace "overcurrent" wherever this word is used to describe fault currents caused by unintentional grounds and short-circuits. Consistent use of these terms makes the requirements easier to understand.

SECTION 427-13. Caution signs located in conspicuous places are necessary to alert workers who have to operate, maintain, and repair the vessel or pipeline equipment.

SECTION 427-14. Heating cables, tapes, strips, and blankets must be secured to the surface being heated. Clips, studs, and bands are used for this purpose.

SECTION 427-15. Where cables and tapes pass over valves, flanges, and other irregular surfaces, it is usually necessary to place a sheet metal backing plate under the heating element to act as a heat sink; otherwise, a hot spot develops. Other types of pipe tracing require splicing nonheating leads between heating sections whenever contact with the pipe surface cannot be maintained.

SECTION 427-16. Nonheating leads are usually spliced to the heating tape where crossing expansion joints.

SECTION 427-17. Manufacturers' instructions must be adhered to when installing resistance heating on flexible pipelines. Here again it may be necessary to bridge the flexible pipe with nonheating wiring.

SECTION 427-18. Field-installed nonheating leads are acceptable provided that the temperature rating of the insulation is as specified. Factory-assembled leads can be shortened provided that the identifying markings are continuous along the length, or if a band is used, it can be removed and relocated after cutting the surplus off. If the leads cannot be shortened without loosing identification, the extra wire must be folded in the junction box.

According to Part (b), metal raceways are required for physical protection of the nonheating leads unless other types of raceways are suitable for this specific application.

Where nonheating leads are used to connect portions of the heating system together, they can be covered with thermal insulation. As mentioned earlier, nonheating leads are often used around valves, strainers, fittings, and other uneven surfaces.

SECTION 427-19. Nonheating interconnections that join portions of the heater elements are permissible without junction boxes provided that the connectors are insulated and then covered with thermal insulation.

Connections to the power supply must be made in a junction box with connectors suitable for the conditions outlined in Section 110-14. The nonheating leads must enter the junction box in metal raceway and the circuit wiring method has to be one of the types mentioned in Cha

SECTION 427-21. All exposed noncurrent-carrying metal parts heating system should be grounded. I say "should" instead of because the wording requires only grounding of exposed metal parts li become energized.

SECTION 427-35. These Sections cover vessels and pipelines heat induction coils operating at line (60-hertz) frequency. If other frequenci employed, Article 665 applies.

SECTION 427-36. Induction coils that operate above 30 volts mu guarded, isolated, or enclosed to protect people in the area from ele shock.

SECTION 427-37. Care must be exercised in the placement of coils to vent circulating currents in adjacent metal equipment, building steel, Where this cannot be avoided, shielding, insulation at joints, or isolatio necessary to prevent circulating currents. Bonding of all metal in the vicir of the coil will prevent arcing between parts because of stray currents.

SECTION 427-45. The single conductor used in a skin-effect currer tracing heating system may carry more current than is shown in the ampaci Tables in Article 310 provided that the conductor is identified for this speci use. Tables 310-16 and 310-18 are not mentioned because the supply volta is not limited to 2000. Since each system is individually designed, overheatin should not occur.

SECTION 427-46. On long runs, pull boxes are necessary. They may be buried in the thermal insulation if their locations are marked on the outer surface of the insulation and their distances from an easily identifiable reference point are shown on drawings.

SECTION 427-47. Since Section 300-20 does not permit a single conductor carrying alternating current in a raceway, this Section is necessary to prevent any misinterpretation.

SECTION 427-48. All lengths of the ferromagnetic tube are welded together and the completed tube is welded to the pipeline. Although grounding at one point might seem adequate, the ferromagnetic envelope must be grounded at each end. If the design requires additional grounding at intermediate points, it shall be done.

SECTION 430-2. If the power conversion equipment for adjustable-speed drives provides overload protection for the motor, additional overload protection is not necessary. Remember that "overload" includes excessive motor current due to overloading and failure to start. This change makes it clear that overload protection in the power conversion unit protects the variable-speed motor from overloading.

SECTION 430-5. Article 675—Electrically Driven or Controlled Irrigation Machines is added to the list because a different method is used to calculate total load current for these multimotor machines.

SECTION 430-6(a). This part is a good example of the proper use of the terms "short-circuit and ground-fault protection" instead of "overcurrent protective devices," and "motor overload" protection instead of "motor running overcurrent" protection.

SECTION 430-10. Part (b) and a wire bending space Table are added for motor controllers. The dimensions shown are a selection of values from Tables 373-6(a) and 373-6(b) and are based on the results of the UL fact-finding investigation of wire bending space in electrical enclosures. Although there is not as much space between the wire terminal and the opposite wall of the controller when compared with Table 373-6(b), motor controller enclosures do not have the same characteristics as cabinets and cutout boxes.

The depth (front to back) of a motor controller enclosure is usually greater than an equivalent-size cabinet, thereby providing more volume and wire bending space. The current rating of the controller and wire terminal size are generally greater than required for the horsepower rating. This usually means that a wire size smaller than the maximum size the lug can accept is installed. In other words, the wire bending space is based on a larger-size wire than actually used. The result is equivalent to having more bending space than required by Table 430-10. Most motor-branch-circuit conductors in small sizes are copper and the wire bending space may be based on aluminum. Controller enclosures are designed for add-on accessories such as auxiliary contacts, control circuit fuses and transformers, pilot lights, selector switches, etc. Seldom are all these components installed; hence more volume (space) is available for bending conductors. Parts of the control mechanism can be removed for better access to the terminals. These are a few of the reasons why slightly less wire bending space is provided in motor controllers than in panelboards and switches. Actually, the differences are not that great. For example, 5 inches is allowed for No. 1/0 in a motor controller while 5.5 inches is required in a panelboard; No. 2/0 and 4/0 are the same in both Tables; and 500 MCM requires 12 inches of space in a motor starter compared to 14 inches in a switch.

If the terminals supplied with the motor controller are changed on the job site for any reason, the replacement terminals must be of a type that have been tested with the controller and identified as suitable by the manufacturer. Substitution of a crimp-type lug for a set-screw lug reduces the wire bending space because crimp-type terminals are longer. And if they are not the same width, the replacement lugs can reduce the clearance between phases.

Table 430-10 is not used for wire bending space in a motor control center. For this equipment Tables 373-6(a) and 373-6(b) are used in accordance with the instructions in Section 373-6.

These changes in wire bending space for motor controllers are a welcome addition and will make many electricians happy.

SECTION 430-12. Volumes for motor terminal housings are increased to reduce the possibility of damaged insulation on the contained conductors. Terminal box dimensions are usually adequate for three-phase induction motors supplied by a three-wire three-phase branch circuit, or where a dual-voltage motor is supplied at the higher voltage. In most other cases it is difficult to get all the wires in the motor terminal box. Since excessive pressure has to be applied, insulation damage results. Where wye–delta starting is used, it is often necessary to remove the motor terminal housing and install a larger junction box. This type of field modification should not be necessary, but it is. Although these increases will probably not solve all field-connection problems, it is a step in the right direction. Let's look at some of the changes.

In the Table with the title "Motors Over 11 Inches in Diameter Alternating-Current Motors," the volume of the terminal housing for a three-phase motor with a full-load current of 100 amperes is increased from 72 cubic inches to 100 cubic inches, an increase of about 40 percent. If we assign dimensions to these volumes, a 72-cubic-inch terminal housing measuring $4 \times 4 \times 4.5$ must now be increased to $4 \times 5 \times 5$ under this revision. These are not typical dimensions of motor terminal housings. They are only used here as an example to show the relative increase in size because of increased volume. Increases in volumes for motor terminal housings on three-phase motors with full-load currents greater than 110 amperes range between 20 and 30 percent.

In the second column of the Table titled "Terminal Box Minimum Dimensions," most of the figures are larger than they were in the past. For example, the minimum dimension was 3.6 inches for a motor with a full-load current of 110 amperes; it is now 4 inches.

The Direct-Current Motor Table is revised to agree with the changes made in the Alternating-Current Motor Table. Notice that the dc motor table is based on not more than 6 wire-to-wire connections and the ac motor table on not more than 12 wire-to-wire terminations. Since there are typically only half as many wires on dc motors, the full-load current value for dc motors

Section 430-32   87

begins with 68 (45 × 1.5) and extends to 900, which happens to be one and one-half times the maximum value (600) shown in the ac motor table.

PART E. Part (e) is new and requires equipment-grounding conductor termination means inside or outside the motor terminal housing. This can be a pigtail, stud, clamp, pressure connector, or other acceptable connection device. Most motors are connected to supply wiring through a short length of flexible metal conduit or liquidtight flexible metal conduit. Since the larger sizes are not recognized as a grounding conductor, an equipment-grounding conductor must be installed in the raceway, or for lengths not exceeding 6 feet, the equipment-grounding conductor can be run on the outside of the conduit provided that it closely follows the raceway surface. Before this revision, it was necessary to ground the motor frame by terminating the equipment-grounding conductor on one of the motor terminal housing screws. It was also common practice to drill a hole in the terminal box and terminate the equipment-grounding conductor with a nut and bolt. Sometimes a liquidtight flexible conduit connector with an exterior grounding terminal was used, but other methods had to be used where the flexible conduit was more than 6 feet long.

Motors that are part of a factory-wired appliance or machine do not need an equipment-grounding terminal if they are grounded by any other effective means, and it is not necessary to connect any field wiring in the motor terminal housing.

SECTION 430-22. This Section requires minimum branch-circuit conductor ampacity to be 125 percent of motor full-load current. The second Exception is added to require branch-circuit conductors supplying a dc motor from an ac source to be increased to 190 percent of motor full-load current for half-wave rectification and 150 percent for full-wave rectification.

PART C. The title of Part C is changed from "Motor Circuit-Conductors" to "Motor and Branch-Circuit Overload Protection," for conformance with the intent of the word "overload" as used in this Article. To refresh your memory, "overload" means any current due to motor overloading and failure to start.

SECTION 430-32. In Part (b), motors of 1 hp or less that are not permanently installed do not need overload protection if not automatically started. Since the words "manually started" were previously used and subject to misinterpretation, they are replaced with "nonautomatically started." The definition of "nonautomatic" is stated in Article 100 as "action requiring personal intervention for its control." It is now clear that a magnetic starter with a start–stop pushbutton is classified as nonautomatic and that the rule does not apply only to so-called "manual" motor starters.

SECTION 430-35. The title of Part (a) is changed from "manually started" to "nonautomatically started" and the parenthetical phrase (including starting with a magnetic starter having pushbutton control) is removed to clarify the intent. This part (a) allows shunting of the motor overload relays during starting of the motor.

SECTION 430-51. The title for Part D was changed from "Motor Branch Circuit Overcurrent Protection" to "Motor Branch Circuit Short-Circuit and Ground-Fault Protection" in the 1975 Edition. This new terminology caused some confusion because "Ground-Fault Protection" is a term used to describe ground-fault circuit-interrupters for personnel protection, and ground-fault protection of equipment is required for 480Y/277-volt electric service for all service disconnecting means rated 1000 amperes or more. Since the change was made to more accurately define the function of motor-branch-circuit overcurrent protection and not to require ground-fault protection for people or equipment, a sentence is added to make it clear that the types of devices mentioned in Sections 210-8 and 230-95 are not required because of the similarity of the language.

SECTION 430-52. Instantaneous-trip circuit breakers or motor short-circuit protectors can be used only as part of a combination motor controller which provides coordinated protection for all components in the circuit. The Exception to this paragraph is rewritten to clearly indicate that the instantaneous-trip setting of the circuit breaker cannot be increased unless the 700 percent of motor full-load current value indicated in Table 430-152 is not sufficient for the motor starting current.

Semiconductors used in adjustable-speed drives are not adequately protected by the overcurrent protective devices listed in Table 430-152. Because fast-acting, energy-limiting fuses are necessary to prevent damage to electronic parts, the last paragraph permits special fuses provided that the type and size are specified at the fuseholders.

SECTION 430-72. Shortly after adoption of the 1978 Edition, a Technical Subcommittee was appointed to clarify, coordinate, and update, if necessary, the rules governing overcurrent protection of motor control circuit conductors. The committee was aptly named—Control Circuit Protection. As a result of this activity, Section 430-72 is completely rewritten and revisions are made to Article 725 dealing with Class 1 and 2 remote-control circuits.

Previously, control circuits were classified as branch circuits, which created problems whenever an attempt was made to apply branch-circuit rules to control wiring. That is changed. Motor Control Circuits are definitely not branch circuits, and this is stated in the second sentence of Part (a).

Previously, only resistance heating was covered. The Article is expanded to include impedance heating, induction heating, and skin-effect current tracing.

SECTION 427-2. A pipeline includes all parts of a closed system used to move fluids; a vessel includes all types of containers used for holding fluids.

Notice that thermal insulation, a moisture barrier, and other nonelectrical components may be required for a properly designed and installed heating system.

The Fine Print Note lists the types of resistance heating elements that are available. Exterior heating elements must be in contact with the heated surface to assure proper heat transfer.

The impedance of the pipeline or vessel is the heating element in this type of heating. Alternating current flows in the vessel or pipeline, causing heating of the metal.

Vessels and pipelines are heated by induction by placing a coil around the object to be heated and applying an ac voltage to the coil.

Skin-effect current tracing (SECT) is accomplished by welding a ferromagnetic conduit to a pipeline, installing a single conductor in the conduit, and connecting it to the far end of the conduit. The ac supply is then connected to the free end of the conductor and the conduit. The skin effect of alternating current causes internal heating of the interior surface of the ferromagnetic tube.

SECTION 427-3. All other appropriate rules in the Code apply unless modified by requirements in this Article. Cord-and-plug-connected heaters are recognized but must meet applicable requirements of Article 422—Appliances.

SECTION 427-4. As with other heating systems, the ampacity of branch-circuit conductors and the rating or setting of the overcurrent devices cannot be less than 125 percent of the total heater load. Where the standard ratings of fuses or circuit breakers do not correspond to the calculated current, the next larger standard size is allowed. Regardless of the heater load current, there is no requirement for subdividing into smaller increments as is necessary for fixed electric space heating.

SECTION 427-10. All parts of the heating system must be suitable for all environmental conditions encountered. Since many of these heating systems are specially designed, it is necessary to follow closely the manufacturer's drawings and installation instructions.

SECTION 427-11. Protection from physical damage is essential to assure reliability and reduce accidents.

SECTION 427-13. Caution signs located in conspicuous places are necessary to alert workers who have to operate, maintain, and repair the vessel or pipeline equipment.

SECTION 427-14. Heating cables, tapes, strips, and blankets must be secured to the surface being heated. Clips, studs, and bands are used for this purpose.

SECTION 427-15. Where cables and tapes pass over valves, flanges, and other irregular surfaces, it is usually necessary to place a sheet metal backing plate under the heating element to act as a heat sink; otherwise, a hot spot develops. Other types of pipe tracing require splicing nonheating leads between heating sections whenever contact with the pipe surface cannot be maintained.

SECTION 427-16. Nonheating leads are usually spliced to the heating tape where crossing expansion joints.

SECTION 427-17. Manufacturers' instructions must be adhered to when installing resistance heating on flexible pipelines. Here again it may be necessary to bridge the flexible pipe with nonheating wiring.

SECTION 427-18. Field-installed nonheating leads are acceptable provided that the temperature rating of the insulation is as specified. Factory-assembled leads can be shortened provided that the identifying markings are continuous along the length, or if a band is used, it can be removed and relocated after cutting the surplus off. If the leads cannot be shortened without loosing identification, the extra wire must be folded in the junction box.

According to Part (b), metal raceways are required for physical protection of the nonheating leads unless other types of raceways are suitable for this specific application.

Where nonheating leads are used to connect portions of the heating system together, they can be covered with thermal insulation. As mentioned earlier, nonheating leads are often used around valves, strainers, fittings, and other uneven surfaces.

SECTION 427-19. Nonheating interconnections that join portions of the heater elements are permissible without junction boxes provided that the connectors are insulated and then covered with thermal insulation.

Connections to the power supply must be made in a junction box with connectors suitable for the conditions outlined in Section 110-14. The non-

Section 427-48

heating leads must enter the junction box in metal raceway and the branch-circuit wiring method has to be one of the types mentioned in Chapter 3.

SECTION 427-21. All exposed noncurrent-carrying metal parts of the heating system should be grounded. I say "should" instead of "shall" because the wording requires only grounding of exposed metal parts likely to become energized.

SECTION 427-35. These Sections cover vessels and pipelines heated by induction coils operating at line (60-hertz) frequency. If other frequencies are employed, Article 665 applies.

SECTION 427-36. Induction coils that operate above 30 volts must be guarded, isolated, or enclosed to protect people in the area from electric shock.

SECTION 427-37. Care must be exercised in the placement of coils to prevent circulating currents in adjacent metal equipment, building steel, etc. Where this cannot be avoided, shielding, insulation at joints, or isolation is necessary to prevent circulating currents. Bonding of all metal in the vicinity of the coil will prevent arcing between parts because of stray currents.

SECTION 427-45. The single conductor used in a skin-effect current-tracing heating system may carry more current than is shown in the ampacity Tables in Article 310 provided that the conductor is identified for this specific use. Tables 310-16 and 310-18 are not mentioned because the supply voltage is not limited to 2000. Since each system is individually designed, overheating should not occur.

SECTION 427-46. On long runs, pull boxes are necessary. They may be buried in the thermal insulation if their locations are marked on the outer surface of the insulation and their distances from an easily identifiable reference point are shown on drawings.

SECTION 427-47. Since Section 300-20 does not permit a single conductor carrying alternating current in a raceway, this Section is necessary to prevent any misinterpretation.

SECTION 427-48. All lengths of the ferromagnetic tube are welded together and the completed tube is welded to the pipeline. Although grounding at one point might seem adequate, the ferromagnetic envelope must be grounded at each end. If the design requires additional grounding at intermediate points, it shall be done.

If the skin-effect current-tracing heating system is supplied from the secondary of a transformer, a separately derived system is created but grounding according to Section 250-26 is not required. Part (d) of this Section permits grounding methods that are prescribed in other portions of the code. In other words, the Fine Print Note refers us to Section 250-26(d), which sends us back to Section 427-48. Catch-22!

SECTION 427-55. The disconnecting means for pipeline and vessel heating systems must be readily accessible to the user. Additionally, the disconnecting means shall be provided with a positive lockout in the off position. There is a difference here compared to the disconnecting means requirements for snow-melting and de-icing heaters in that a lockout feature is not necessary for snow-melting and de-icing.

SECTION 427-56. This Section details various constructions of temperature controllers that are or are not acceptable as disconnecting means for pipeline and vessel heating. Since they are identical to the rules in Section 426-51, no more explanation is necessary.

SECTION 427-57. The rating of overcurrent protective devices is based on the ampacity of the branch-circuit conductors supplying the heating equipment.

Article 430
Motors, Motor Circuits, and Controllers

DIAGRAM 430-1. Editorial revisions of the 1975 NEC clarified the terms overcurrent, overload, and short-circuit, and ground-fault protection. "Overcurrent" in this Article and Article 440 means current resulting from overload, short-circuit, and ground-fault. "Overload" means excessive motor current due to overloading and failure to start. "Short-circuit and ground-fault protection" refers to the motor-branch-circuit protective devices. These terms appear in the Diagram. Additional revisions are made in these Articles by substituting "motor overload" for "motor running overcurrent" and "motor running overloads." Also, "short-circuit and ground-fault" replace "overcurrent" wherever this word is used to describe fault currents caused by unintentional grounds and short-circuits. Consistent use of these terms makes the requirements easier to understand.

SECTION 430-2. If the power conversion equipment for adjustable-speed drives provides overload protection for the motor, additional overload protection is not necessary. Remember that "overload" includes excessive motor current due to overloading and failure to start. This change makes it clear that overload protection in the power conversion unit protects the variable-speed motor from overloading.

SECTION 430-5. Article 675—Electrically Driven or Controlled Irrigation Machines is added to the list because a different method is used to calculate total load current for these multimotor machines.

SECTION 430-6(a). This part is a good example of the proper use of the terms "short-circuit and ground-fault protection" instead of "overcurrent protective devices," and "motor overload" protection instead of "motor running overcurrent" protection.

SECTION 430-10. Part (b) and a wire bending space Table are added for motor controllers. The dimensions shown are a selection of values from Tables 373-6(a) and 373-6(b) and are based on the results of the UL fact-finding investigation of wire bending space in electrical enclosures. Although there is not as much space between the wire terminal and the opposite wall of the controller when compared with Table 373-6(b), motor controller enclosures do not have the same characteristics as cabinets and cutout boxes.

The depth (front to back) of a motor controller enclosure is usually greater than an equivalent-size cabinet, thereby providing more volume and wire bending space. The current rating of the controller and wire terminal size are generally greater than required for the horsepower rating. This usually means that a wire size smaller than the maximum size the lug can accept is installed. In other words, the wire bending space is based on a larger-size wire than actually used. The result is equivalent to having more bending space than required by Table 430-10. Most motor-branch-circuit conductors in small sizes are copper and the wire bending space may be based on aluminum. Controller enclosures are designed for add-on accessories such as auxiliary contacts, control circuit fuses and transformers, pilot lights, selector switches, etc. Seldom are all these components installed; hence more volume (space) is available for bending conductors. Parts of the control mechanism can be removed for better access to the terminals. These are a few of the reasons why slightly less wire bending space is provided in motor controllers than in panelboards and switches. Actually, the differences are not that great. For example, 5 inches is allowed for No. 1/0 in a motor controller while 5.5 inches is required in a panelboard; No. 2/0 and 4/0 are the same in both Tables; and 500 MCM requires 12 inches of space in a motor starter compared to 14 inches in a switch.

If the terminals supplied with the motor controller are changed on the job site for any reason, the replacement terminals must be of a type that have been tested with the controller and identified as suitable by the manufacturer. Substitution of a crimp-type lug for a set-screw lug reduces the wire bending space because crimp-type terminals are longer. And if they are not the same width, the replacement lugs can reduce the clearance between phases.

Table 430-10 is not used for wire bending space in a motor control center. For this equipment Tables 373-6(a) and 373-6(b) are used in accordance with the instructions in Section 373-6.

These changes in wire bending space for motor controllers are a welcome addition and will make many electricians happy.

SECTION 430-12. Volumes for motor terminal housings are increased to reduce the possibility of damaged insulation on the contained conductors. Terminal box dimensions are usually adequate for three-phase induction motors supplied by a three-wire three-phase branch circuit, or where a dual-voltage motor is supplied at the higher voltage. In most other cases it is difficult to get all the wires in the motor terminal box. Since excessive pressure has to be applied, insulation damage results. Where wye–delta starting is used, it is often necessary to remove the motor terminal housing and install a larger junction box. This type of field modification should not be necessary, but it is. Although these increases will probably not solve all field-connection problems, it is a step in the right direction. Let's look at some of the changes.

In the Table with the title "Motors Over 11 Inches in Diameter Alternating-Current Motors," the volume of the terminal housing for a three-phase motor with a full-load current of 100 amperes is increased from 72 cubic inches to 100 cubic inches, an increase of about 40 percent. If we assign dimensions to these volumes, a 72-cubic-inch terminal housing measuring $4 \times 4 \times 4.5$ must now be increased to $4 \times 5 \times 5$ under this revision. These are not typical dimensions of motor terminal housings. They are only used here as an example to show the relative increase in size because of increased volume. Increases in volumes for motor terminal housings on three-phase motors with full-load currents greater than 110 amperes range between 20 and 30 percent.

In the second column of the Table titled "Terminal Box Minimum Dimensions," most of the figures are larger than they were in the past. For example, the minimum dimension was 3.6 inches for a motor with a full-load current of 110 amperes; it is now 4 inches.

The Direct-Current Motor Table is revised to agree with the changes made in the Alternating-Current Motor Table. Notice that the dc motor table is based on not more than 6 wire-to-wire connections and the ac motor table on not more than 12 wire-to-wire terminations. Since there are typically only half as many wires on dc motors, the full-load current value for dc motors

Section 430-72

Control circuits derived from the load side of motor-branch-circuit short-circuit and ground-fault protective devices are covered by rules in this Section, while motor control circuits supplied from other sources must be installed according to applicable rules in Article 725, particularly Sections 725-12 for Class 1 circuits and 725-35 for Class 2 circuits.

Since line-voltage motor control circuits tapped from the load side of the motor-branch-circuit protective devices are no longer considered branch circuits, glass fuses or other supplementary overcurrent devices can be used where conditions warrant.

The basic rules for overcurrent protection of control circuit wiring are given in Part (b), but there are Exceptions. For a good understanding of the requirements, an explanation of the basic rules, then of the Exceptions, is given, to point out the allowable variations.

Control conductors No. 12 and larger must be protected according to their ampacities, listed in Tables 310-16 through 310-19. Regardless of the number of conductors in a raceway, derating is not required. This is because all conductors that may occupy the same raceway do not generally carry their rated current continuously. Since the Scope Section of Article 310 exempts internal wiring of motor controllers from the requirements of that Article, and control circuit protection applies to both internal and external control circuits, it is necessary to point out that Section 310-1 does not apply. This avoids a conflict. Control circuit conductors of Nos. 18, 16, and 14 require overcurrent protection not greater than 20 amperes. Now let's look at the four Exceptions that modify these simple, basic rules. These Exceptions can be viewed as "if–then" situations.

Inside a motor starter the control conductors for cover-mounted on-off-automatic selector switches or start–stop pushbuttons are wired with Nos. 18, 16, or 14 stranded conductors. Where the control circuit wiring does not extend beyond the motor controller enclosure, they are considered as adequately protected if the branch-circuit-overcurrent protective devices do not exceed 400 percent of the ampacity of the control wiring. Because Table 310-17 does not list ampacities for Nos. 18 and 16, the maximum branch-circuit-overcurrent devices are listed as 25 amperes for No. 18, and 40 amperes for No. 16. These values are based on an assumed ampacity (see Table 400-5) of 7 amperes for No. 18 and 10 amperes for No. 16. For No. 14 and larger conductors, the 60°C column in Table 310-17 is used to obtain the ampacity. Therefore, the maximum motor-branch-circuit short-circuit and ground-fault protective devices for No. 14 is 100 amperes, and for No. 12 it is 120 amperes. It may look as if these figures are incorrect, but they are not. The ampacity of No. 14 in Table 310-17 is 25 amperes and No. 12 has an ampacity of 30 amperes.

If the branch-circuit protective devices are set higher than the maximum values permitted for the installed control circuit wires, internal overcurrent protection is necessary. Some manufacturers install and connect control circuit overcurrent protection devices at the factory, whereas others have them available as an optional add-on feature.

Where the control wiring extends beyond the controller enclosure and is No. 14 or larger, additional overcurrent protection for this wiring is unnecessary if the motor-branch-circuit overcurrent protection is not greater than 300 percent of the ampacity of the control wires. The ampacity is obtained from Table 310-16 using the 60°C column. Note that the ampacity for external control wiring is determined from Table 310-16 and for internal control wires Table 310-17 is specified. Suppose that we are using No. 14 control wires for limit switches and the controller has a start–stop pushbutton mounted on the cover, factory-wired with No. 16 conductors. According to Exception No. 1, the No. 16 must have protection not in excess of 40 amperes, and the external wiring to the limit switches is limited to 60 amperes. Therefore, the motor-branch-circuit short-circuit and ground-fault protection cannot exceed 40 amperes, or if it does, control circuit overcurrent protection is required.

Because of the wording in the beginning of Exception No. 2 you may ask: Can Nos. 18 and 16 be used for external control wiring or is the minimum size limited to No. 14? It is now time to go back and review paragraph (b)(2), which is the basic rule. Here we find that Nos. 18 and 16 can be used for control wiring without restrictions as to location or length as long as the overcurrent protection does not exceed 20 amperes. Therefore, the answer is yes—these smaller sizes can be used. But caution should be exercised because a long run of No. 18 to a remote control pushbutton or indicator light may have enough impedance to prevent opening of the 20-ampere overcurrent device should a fault develop at the far end.

Where the control circuit is obtained from a transformer, Exception No. 3 applies. If the secondary is two-wire and the primary overcurrent protection is in accordance with Section 450-3, the secondary control conductors might not need additional overcurrent protection. Assume a 1200-VA single-phase 480/120-volt transformer with 3-ampere primary fuses. The maximum secondary current under these conditions is 12 amperes. Therefore, Nos. 18, 16 or 14 can be used for control wiring without overcurrent protection on the secondary. On the other hand, control power for a motor control center may be obtained from a 15-kVA 480/120-volt single-phase transformer protected on the primary side with a 40-ampere two-pole circuit breaker. The maximum secondary current now is (4×40) 160 amperes and overcurrent protection must be provided for all control circuit wiring. This is usually accomplished by providing a control circuit panelboard in the motor control center.

Control circuit protection may be omitted by Exception No. 4 if opening of the circuit creates a greater hazard than would result from damage to the control circuit wiring. Fire pumps and motors used in a continuous process requiring orderly shutdown are examples.

For control circuit transformers the basic rule is that the transformer primary windings be protected from overcurrent as outlined in Article 450.

Primary overcurrent protection can be eliminated according to the first Exception if the transformer is in the motor controller enclosure and secondary protection not exceeding 200 percent of rated secondary current is provided. The primary winding of the control transformer is usually tapped from the line terminals of the controller, and the secondary overcurrent protection is installed and connected in the factory.

Power-limited Class 1, 2, and 3 control circuits are governed by the rules in Article 725. Overcurrent protection for Class 1 control conductors is not necessary if the power-limited Class 1 transformer is protected on the supply side by overcurrent devices sized according to Section 450-3 and control wiring overcurrent protection conforms with Section 725-12. For Class 2 and 3 circuits, the transformers must be protected by overcurrent devices not exceeding 20 amperes.

Exception No. 4 allows omission of overcurrent protection for control wiring where fire pumps and similar types of motor-operated equipment are involved.

SECTION 430-109. Exception No. 5 recognized a cord and plug as a disconnecting means for motors where a flexible cord connection is allowed under one of the permitted uses listed in Section 400-7. The Exception formerly stated that portable motors could be cord-and-plug-connected, but "portable" was not defined. Large motors that are so heavy they have to be moved with a crane were called portable. A portable building is another example of how big a portable object can be.

Since the Exception was being misapplied, it is rewritten to more fully explain the intent. Cord-and-plug-connected appliances are acceptable if the attachment plugs and connectors have adequate interrupting capacity, are noninterchangeable, and meet the other requirements of Section 422-22. Room air conditioners with ratings not over 40 amperes and 250 volts single-phase can also be cord-and-plug-connected if the requirements of Section 440-62 are satisfied. For any other cord-and-plug-connected motors, the plug and receptacle or cord connector must be horsepower-rated. Generally, attachment plugs and receptacles rated 60 amperes or less only have to interrupt 150 percent of their rating. Attachment plug–receptacle combinations rated above 60 amperes are marked "For disconnect use only," meaning they have not been tested for current interrupting. Because of the nonavailability of suitable receptacles and attachment caps, the revision has the effect of limiting cord-and-plug connections to small motor-operated appliances, as originally intended.

SECTION 430-110. This Section contains detailed information on how to size the disconnecting means for a single motor, a group of motors, or motors combined with other leads. Part (c) covers sizing of the disconnecting means where the motor is connected to the same circuit that supplies resistance heating elements. This method works fine where the heating load is small as compared to the motor load, but the opposite is not true. For this reason the Exception is added. To see how this works, let's make some assumptions. Suppose that we have to size a disconnecting means for a 5-hp three-phase 230-volt blower on the same circuit with a 25-kW three-phase duct heater. The full-load current of the motor is 15 amperes and the locked-rotor current is 90 amperes. According to paragraph (c)(1), the total current is obtained for motor-running conditions and locked-rotor conditions by adding the heating load to these two values of motor currents. These results are used to determine the rating of the disconnecting means.

Under normal operating conditions the full-load current of the motor is 15 amperes. This is added to the heater full-load current, which is approximately 63 amperes, for a total of 78 amperes. Now turning to Table 430-150, we move down the amperes column for 230 volts until we find a value equal to or greater than 78 amperes. At this point (80 amperes) read the horsepower figure horizontally to the left, which is 30. Therefore, for normal operation a disconnect with a 30-hp rating is required, but this is not the final answer. The same procedure must be followed to find the minimum horsepower rating under locked-motor conditions by adding 90 amperes (locked-rotor current) to the heater full-load current. Table 430-151 is used to obtain equivalent horsepower for a total current of (90 + 63) 153 amperes. Proceeding down the 230-volt two- or three-phase column as before, we stop at 168 amperes and read 10 under the maximum horsepower rating column on the right. Since the horsepower rating of the switch must be the larger of the two values obtained, a 30-hp rated switch is required, or this is the way it was before the Exception was added.

Actually, the Exception is added because of the unnecessarily large disconnecting means required where the motor load is a small portion of the total load. Applying the Exception to this combination allows a switch horsepower rating not lower than the motor load and an ampere rating equal to or greater than the locked-rotor current of the motor plus the heater load. Therefore, a 200-ampere switch with at least a 5-hp rating is required as compared to a 400-ampere 30-hp rated switch under the former method. Since a 200-ampere fused safety switch has a standard horsepower rating of 25, it certainly satisfies that part of the Exception which requires the horsepower rating of the switch to be at least equal to the total motor load expressed in horsepower. I must point out that most horsepower rated switches have dual ratings and the ratings mentioned previously are standard. Whenever the maximum horsepower rating is applied, time-delay fuses are usually necessary. This is mentioned so that you are not mislead by my answer that 400-ampere switch is required to obtain a horsepower rating of 30 or more, since most 240-volt three-phase 200-ampere heavy-duty safety switches have a standard rating of 25 hp and a maximum rating of 60 hp.

Since this new Exception recognizes an alternative method for calculating ampere and horsepower ratings of disconnecting means where resistance heating is part of the circuit load, both methods should be tried; then use the results that allow a lower rating for the switch.

SECTION 430-113. This Section requires disconnecting means for each supply source where a motor or motor-operated equipment is supplied by more than one energy source. Exception No. 2 now exempts Class 2 remote-control circuits which are ungrounded and operate at 30 volts or less. Since these circuits do not ordinarily create a shock or fire hazard, there is no reason to disconnect them while working on equipment in which they are installed.

SECTION 430-145. A Fine Print Note is added to reference the motor terminal housing grounding means required in Section 430-12(e).

TABLE 430-149. Full-load currents for sychronous-type unity-power-factor motors are removed from this Table. The submitter of this proposal suggested that the voltages 220, 440, and 550 shown in the columns of the Table be increased to 230, 460, and 575. As an alternative the proposer recommended deletion of all information dealing with two-phase synchronous motors because there is no longer a need for this information.

TABLE 430-151. The title is changed to indicate that the table is to be used only for conversion to equivalent horsepower where the locked-rotor current is stamped on the nameplate. Usually, nameplates on hermetic refrigerant motor-compressors include voltage, number of phases, rated-load current, and locked-rotor current. Since a horsepower rating is not included, it is necessary to convert to equivalent horsepower by using appropriate tables. Table 430-151 is used to obtain equivalent horsepower under locked-rotor condition. On standard motors the rated horsepower is included along with other information on the nameplate. For these motors locked-rotor current is determined from the Code Letter and Conversion Table 430-7(b). Depending on the Code Letter, the locked-rotor current can be greater or less than the values shown in Table 430-151, which are based on six times motor full-load current.

The statement "For use only with Sections 430-110, 440-12, and 440-41" further clarifies the restricted application of the Table.

Article 440
Air-Conditioning
and Refrigerating Equipment

SECTION 440-12. The rating and interrupting capacity of the disconnecting means for hermetic refrigerant motor-compressors with or without other motors and loads on the same branch circuit are determined by the rules in this Section. For a fused switch the rating refers to the ampere rating and the interrupting capacity to the horsepower rating.

Part (b) deals with combination loads consisting of other motors and/or resistance loads that are connected to the same branch circuit that supplies the hermetic refrigerant motor-compressor. Generally, the full-load currents of motors and heaters are added to the compressor rated-load current or branch-circuit selection current (use the larger of the two) to obtain a total load current. This figure is used to determine the horsepower rating of the switch by using whichever Table (430-148, -149, or -150) is appropriate. Next, the locked-rotor currents of all motors plus the locked-rotor current of the compressor are added to the full-load current of the resistance heater, if present, to get a total locked-rotor current. This figure is used with Table 430-151 to determine the horsepower rating under locked-rotor conditions. Finally, the horsepower rating under full-load conditions is compared with the value obtained under locked-rotor conditions and the larger of the two is the minimum horsepower rating of the switch.

The Exception is new and recognizes a different method for calculating the horsepower rating of the disconnecting means. This Exception is similar to the one added to Section 430-110, which was explained with an example.

The important difference between the basic rule and the Exception is that an equivalent horsepower rating is not calculated under full-load conditions. The switch only has to have a horsepower rating equal to the total locked-rotor currents of the motors and compressor, provided that it has an ampere rating equal to this locked-rotor current plus the full-load current of the resistance heating.

Article 445
Generators

SECTION 445-5. The ampacity of the conductors connected to the output terminals of a generator must be 115 percent of nameplate ampere rating. Previously, this requirement did not take into account the reduction in neutral conductor ampacity where some loads are not connected phase-to-neutral, and the demand factor of 70 percent that can be applied to that portion of the neutral load exceeding 200 amperes. These are allowable deductions according to Section 220-22 and can now be applied to the generator feeder neutral provided that the reduced size is not smaller than listed in Table 250-94, where the neutral must carry ground-fault current.

Since it is not clear why a reference is made to Section 250-23(b), which has to do with providing a grounded conductor for a service that is supplied by a grounded system, an explanation is in order. The neutral conductor of a generator does not have to be grounded at its terminal if the neutral is connected to another system that has a grounded neutral. Section 250-5 (d) does not require grounding of the neutral of a separately derived system where the separately derived system neutral is interconnected with the grounded neutral of the utility supply. An example of this arrangement is an emergency generator connected to an automatic transfer switch where the normal source neutral and the generator neutral are terminated on a neutral bus in the transfer switch. The generator neutral is actually grounded at the service equipment from which the normal supply to the transfer switch originates. Therefore, the generator neutral is grounded, and the grounding electrode at the service serves a dual function. It is used for grounding the neutral of the service and the generator. But what happens when a phase-to-ground fault occurs on the load side of the transfer switch when the generator is operating? The fault current flows on a phase conductor of the generator to the point of failure, then returns on the equipment-grounding conductor, which may be a metal raceway, all the way back to the service equipment enclosure, then onto the bonding jumper, from the bonding jumper to the neutral back to the transfer switch, where it continues on the generator neutral conductor to the generator terminal. Now, the ground-return path is complete and the generator overcurrent protection opens the circuit. Notice that during this sequence the generator neutral carries ground-fault current. For this reason the generator neutral cannot be decreased below the size specified in Section 250-23(b).

Article 450
Transformers
and Transformer Vaults

SECTION 450-2. Some Sections of this Article allow transformer installations indoors provided that the room or area is of fire-resistant construction. Since "fire resistant" was not defined, varying interpretations resulted. The new paragraph requires fire-resistant construction to have a fire rating of 1 hour. The Fine Print Note lists standards for making this evaluation.

TABLE 450-3(a)(2). The sizes of the primary and secondary fuses are increased. Where the impedance is not more than 6 percent and the secondary is over 600 volts, the fuse rating for the secondary is increased from

150 to 250 percent. For transformers with impedance above 6 and not greater than 10 percent, the primary fuses are increased from 200 to 300 percent and the secondary (over 600 volts) fuses from 125 to 225 percent. These changes should not have any adverse effects, because a transformer with its large thermal capacity can handle the overloading until the fuses open. Large inrush currents occur when a transformer is energized and have caused fuse melting. These higher percentages should eliminate this problem.

SECTION 450-8. The requirement for ventilation is now specific, since the revision makes it clear that there must be enough air movement to keep the ambient temperature below the value marked on the transformer nameplate.

SECTION 450-22. Dry-type transformers exceeding 112½ kVA installed outdoors must be kept 12 inches away from combustible building surfaces. Since dry-type transformers of this same rating must be separated from combustible material when installed indoors, similar precautions must be taken for outdoor applications.

SECTION 450-23. With the demise of Askarel because of regulations placed on polychlorinated biphenyls (PCBs) by the Environmental Protection Agency, substitute transformer insulating fluids had to be developed. The 1978 Edition of the NEC recognized a high-fire-point liquid but did not provide specific rules for its use. In fact, Askarel-filled transformers were being replaced with high-fire-point-liquid-insulated transformers without regard to increased fire risks. This is now changed.

First, the liquid has to be listed by a recognized independent testing laboratory. The listing includes fire point of the liquid, heat-release rate, etc. Transformers containing listed less-flammable liquids can be installed in noncombustible areas of noncombustible buildings provided that a curb, metal pan, or dike surrounds the transformer to contain the liquid should the transformer case leak or rupture. Clearances must be maintained according to the parameters assigned to each type of listed liquid. The heat-release rate is considered in assigning clearances and is determined from the heat of combustion and rate of burning. Where the transformer is not installed according to the specifications for the fluid involved, the area must be protected by sprinklers, water spray, carbon dioxide (CO_2) or another type of automatic fire extinguishing system.

Transformers rated over 35 kV have to be installed in a vault. This voltage limitation applies to all transformers regardless of the cooling medium.

Where transformers containing a high-fire-point liquid are installed outdoors, they are treated the same as oil-filled transformers. At first glance this requirement may seem overly restrictive because some of the listed less-flammable liquids are not nearly as much of a fire hazard as transformer oil,

Section 460-7

but on the other hand, clearances from combustible materials for oil-filled transformers are not hard to satisfy.

Although there are other burning characteristics of less-flammable liquids that require consideration for proper application, the only specification in the NEC for this material is that it have a fire point not less than 300°C.

SECTION 450-24. This is a new Section dealing with nonflammable fluid-insulated transformers. There are no special restrictions on clearances from combustible or noncombustible construction either indoors or outdoors. However, if rated over 35 kV, they must be installed in a vault.

SECTION 450-42. Automatic fire extinguishing systems in transformer vaults allow construction of the walls, ceiling, and floor to be reduced from 3 hours to 1 hour. Since Halon has been used successfully as an extinguishing agent, it is now recognized.

SECTION 450-43. Halon is added in this Exception, too.

SECTION 450-45. Because of unusual circumstances, it is sometimes necessary to provide ventilation openings on inside walls of transformer vaults. When this happens, louvered openings are inside the building and the fire-rated wall construction is lost. Automatic-closing fire dampers with a 1 ½-hour fire rating are now required in these openings because of a revision to Part (e).

Article 460
Capacitors

SECTION 460-7. Revisions are made to clarify the intent and prevent the use of capacitors for some motor applications. The former wording in Part (a) gave the impression that the motor power factor was being corrected. Since capacitors do not correct motor power factor, the last part of the sentence is changed to state that the power factor of the motor branch circuit is raised.

The Exception permits an increase (maybe) in kVAR up to 50 percent of rated input for motors up to 50 hp operating at 600 volts or less. This allows you to choose capacitors based on no-load power factor or 50 percent of motor kVA, whichever gives the most improvement in power factor.

Part (b) was formerly part of an Exception that is rewritten to prohibit capacitors on circuits supplying motors that are subject to unusual switching operations.

SECTION 460-8. The disconnecting means for a capacitor bank is now required to open all ungrounded conductors simultaneously. Previously single-pole switching was acceptable. The main reason for this change is improved safety to operating personnel.

Article 470
Resistors and Reactors

There are no substantial changes in this Article.

Article 480
Storage Batteries

There are no substantial changes in this Article.

FIVE

Special Occupancies

FIVE

Special Occupancies

Article 500
Hazardous (Classified) Locations

As a result of a proposal submitted by the American Petroleum Institute, the word "hazardous" is replaced with "classified" or other appropriate phrases in Articles 500 through Article 516. Where "hazardous" is not removed, "(classified)" is added. The title of this Article is one example. These substitutions do not change the intent. The word "hazard" generally means a continuous unsafe condition, which is not necessarily the case if the area is properly classified and electrical equipment is installed and maintained according to the NEC.

SECTION 500-1. The Scope is made more definitive by pointing out the types of fire or explosive hazards covered by Articles 501, 502, and 503.

Some Fine Print Notes are revised, such as the addition of an installation standard for intrinsically safe instruments. NFPA Standard 45-1975 covering Fire Protection of Laboratories Using Chemicals is added to provide guidance in classifying these areas. Chemistry laboratories are usually considered as ordinary locations, but the standard should be reviewed before making this determination.

SECTION 500-2. Although some explosive substances are added to Table 500-2, it is still not all-inclusive because some hazardous materials are omitted and others have not been classified. So that code users are aware of this situation, the second sentence is added to the fourth Fine Print Note.

Values of resistivity are now assigned to dusts. For Group E the value is 10^2 ohm-cm or less. Generally, these dusts are considered electrically conductive. Group F dusts have a resistivity between 10^2 and 10^8 ohm-cm. All dusts under the Group G category have a resistivity greater than 10^8 ohm-cm and are regarded as electrically nonconductive. Although specific values of resistivity are assigned to each group, you should be aware that various test procedures do not give consistent results.

TABLE 500-2. The list of chemicals is expanded to include: arsine and propylnitrate in Group B: ethyl mercaptan, ethyl sulfide, and hydrogen cyanide in Group C; ethylene glycol monomethyl ether in Group D. Sec-butyl alcohol is deleted because it is the same material as 2-butanol (secondary butyl alcohol), which appears in the Table under Group D.

Part (a) of Section 500-2 is revised to make it clear that electrical equipment must not only be approved for the class of location, but also for the explosive, combustible, or ignitible properties of the specific gas, vapor, dust, fiber, or flyings that are present. This language makes it clear that special precautions must be observed for some chemical atmospheres beyond those required for the group classification in which it belongs. An example of additional safeguards is mentioned in the Fine Print Note under Group G atmospheres for carbon disulfide.

Part (b) is revised to make it clear that equipment does not have to operate in a 40°C ambient. Previous wording implied that electrical equipment had to be installed in a 40°C ambient, whereas the intent was to have the operating temperature marked on the equipment based on a 40°C ambient.

Exceptions Nos. 3 and 4 point out that equipment (other than lighting fixtures) does not have to be specifically listed for Class, Group, Division, or operating temperature in Class I and II Division 2, and Class III locations. In these locations (Class II and III) the equipment must be dust-tight, but special markings are not required. Also, equipment—not including lighting fixtures—suitable for installation in a Class I, Division 2 location does not require Class, Group etc., marking on the enclosure.

SECTION 500-5. The Fine Print Note dealing with open bins, hoppers, blenders, mixers, etc., is removed, since modern equipment is not operated in this fashion.

A new sentence is added which defines electrically conductive dusts as those having a resistivity of less than 10^5 ohm-cm. When compared with the resistivity limits mentioned in Section 500-2, it becomes obvious that some Group F dusts are electrically conductive and others are not. Any dusts in this Group with a resistivity between 10^2 and 10^5 ohm-cm are electrically conductive, and dusts with a resistivity above 10^5 to 10^8 ohm-cm are considered electrically nonconductive.

Part (b) is reworded to state that Class II, Division 2 locations are those which do not have dust suspended in the atmosphere in sufficient quantities to produce an explosive or ignitable mixture under normal operating conditions. All electrical equipment must be capable of operating safely under these conditions. In the event of an equipment (not electrical) malfunction, the electrical equipment must still function as intended, but dust ignition may result from abnormal operation or failure of electrical equipment.

SECTION 500-6. In the first Fine Print Note, "sawmills" are added to the list of locations that belong under Class III. The second Fine Print Note includes sawdust and woodchips to correlate with the addition of sawmills. Although woodworking plants were previously mentioned—now changed to woodworking locations—this phrase generally applies to operations where wood purchased from a sawmill is the raw material that is processed into finished products such as doors, frames, cabinets, furniture, etc. Since sawdust and wood chips accumulate in sawmills, they are subject to the same hazards as a wood working location.

Article 501
Class I Locations

SECTION 501-5. Seals are required in each conduit entering or leaving an enclosure containing devices that may produce arcs, sparks, or high temperatures, where the equipment is installed in a Class I, Division 1 location. The Exception removes the requirement for seals on all conduits up to and including 1 ½ inches where contacts are either hermetically sealed or immersed in oil. Mercury-tube switches are one type of hermetically sealed arcing devices. The conduit size is limited because all conduits 2 inches or larger have to be sealed at enclosures housing splices or taps. For oil-immersed contacts, the level must be at least 2 inches above contacts that make and break power circuits and 1 inch for control circuits.

Sealing compound and the method of application must assure against the passage of gas or vapor through the seal fitting. This change appears in Part (c)(2) and replaces "approved for the purpose." Conductors should be separated from each other in the fitting so that compound completely surrounds each wire.

A Fine Print Note is added at the end of Part (e)(2) to point out that the minimum leakage for gas/vapor-tight cables does not include any leakage that may occur between the strands of individual conductors that are part of the cable.

SECTION 501-6. If internal fuses are used in lighting fixtures they must be of the cartridge type, filled with noncombustible granular material. Quartz powder is one type of filler commonly used. This additional requirement is Part (b)(5).

SECTION 501-11. This Section recognizes flexible cords in Class I, Divisions 1 and 2 for connection of portable utilitization equipment. The second paragraph permits flexible cord on a submersible pump where provisions are made to remove the pump without entering the wet-well of a wastewater pumping station.

SECTION 501-16. All electrical equipment installed in Class I, Divisions 1 and 2 areas must be bonded to the service equipment as indicated in Part (b). Methods of bonding are detailed in Section 250-79 and are applicable. This Section permits external bonding under limited conditions where the bonding jumper is routed with the raceway. The revision here says that the bonding jumper must be installed in parallel with the conduit. Although the language is different, the intent is the same.

Revisions are made in Part (c) dealing with lightning protection to agree with the rewrite of Article 280—Surge Arresters. To refresh your memory, Article 280 covers selection, installation, and connection of surge arresters. Methods of grounding are also included, and Article 250 is referenced for grounding connections. The requirement that surge arresters be installed in an enclosure suitable for the Group and Class I location is new.

Part (d) is revised to recognize a grounding method detailed in Part (e). Revisions in Part (d) include references to particular Sections in Article 250 which require that the grounded conductor be run to the service equipment where the ac supply includes a grounded conductor. The new Exception removes the requirement for a grounded conductor to be run with the circuit conductor where the system grounded conductor is not used and grounding is accomplished in accordance with Part (e).

Where the system-grounded conductor is not used as a circuit conductor and objectionable current flow over the grounding conductors exists because of multiple grounding, the grounded conductor does not have to be run to the service equipment. However, the system-grounded conductor must be grounded at the transformer. With this arrangement there is only one ground on the system. To assure a low-impedance ground path should a ground-fault develop on the load side of the service equipment, a grounding conductor sized according to Section 250-23(b) must be installed with the service conductors. This conductor is connected to the grounded terminal of the transformer and the equipment grounding bus in the service enclosure. At this point the equipment grounding conductor is connected to a grounding electrode through a grounding electrode conductor.

Article 502
Class II Locations

SECTION 502-1. The last paragraph repeats information contained in Section 500-5(a) and assigns a value of 10^5 ohm-cm as the upper limit of resistivity for electrically conductive dusts.

SECTION 502-3. This Section is revised by deleting information now included in Article 280—Surge Arresters. Also, a disconnecting means is required to disconnect fuses protecting surge-protective capacitors.

The requirement that surge arresters be placed in Class II enclosures suitable for the location is new.

SECTION 502-5. Any raceways connecting dust-ignition-proof enclosures to ordinary location types of enclosures in unclassified areas do not have to be sealed, even though the raceways running between the two enclosures are less then the lengths specified in the first paragraph. For example, if a general-purpose switch in an unclassified area feeds a dust-ignition-proof junction box in a Class II, Division 1 area on the other side of the wall on which the switch is mounted, a seal fitting is not necessary even though the horizontal conduit connecting the two enclosures is less than 10 feet long.

SECTION 502-6. Part (b) is revised to state that fuses, switches, relays, etc., installed in Class II, Division 2 locations must be mounted in dust-tight enclosures.

Since "dust-tight" is defined as "so constructed that dust will not enter the enclosing case under specified test conditions," it is not necessary to provide any more detail. The previous edition said that these enclosures had to comply with Section 502-6(a)(2).

Dust-tight enclosures are now required for switching mechanisms instead of having to comply with Section 502-6(a)(2).

SECTION 502-8. Now, totally enclosed nonventilated and totally enclosed fan-cooled motors and generators are acceptable for use in Class II, Division 2 locations.

In part (b) of the Exception "dust-tight housings" replaces "tight metal housings."

SECTION 502-16. It is necessary to ground and bond all exposed non-current-carrying metal parts of electrical equipment installed in Class II, Division 1 and 2 areas. When bonding flexible metal conduit, the jumper may be inside or outside the raceway as permitted by Section 250-79.

Part (c) is revised and requires a grounded service conductor in the service raceway if the ac supply system is solidly grounded. This is required by Section 250-23(b) and applies to all services. The old rule only required bonding the service raceway to the grounded service conductor and grounding conductor of the raceway system. Actually, this revision does not change the intent, but the Exception and Part (d) do.

The Exception says that the grounded service conductor does not have to be installed in the service raceway if it is not used as a circuit conductor for any loads supplied from the service. However, grounding must be provided as outlined in Part (d). Since this method of grounding was discussed in detail under Section 501-16(e), additional explanation is not necessary.

Article 503
Class III Locations

SECTION 503-9. A requirement that boxes used for the support of lighting fixtures be approved for the purpose is deleted because Section 503-3(a)(1) provides construction details for boxes and fittings, while Parts (a) and (c) describe supporting methods for fixed and pendant lighting fixtures. In other words, the requirement was redundant.

Article 510
Hazardous (Classified) Locations Specific

There are no changes in this Article.

Article 511
Commercial Garages, Repair and Storage

In many parts of this Article (classified) is substituted for "hazardous," "Class I" is another alternative for "hazardous"; also, "flammable vapors" is used instead of "hazardous vapors."

SECTION 511-5. In Part (c), the title is now Grounded Conductor and "identified conductor" is replaced with "grounded conductor" in two places in the text. Because of the new definition for "identified," which means electrical equipment suitable for a specific use, purpose, function, etc., this word can no longer be used as a substitute for grounded conductor.

Article 513
Aircraft Hangers

Here again, there are numerous changes from "hazardous" to "Class I," but none of these alter the rules as previously existing.

Article 514
Gasoline Dispensing and Service Stations

SECTION 514-2. Distances that Class I, Divisions 1 and 2 locations extend beyond various types of equipment in a service station are now in Table form. This information, with slight modification, is taken from NFPA 30-

Section 514-2

1977, *Flammable and Combustible Liquids Code*. The revisions remove any conflicts that may have existed between the two Codes and adds classified areas not previously covered by the NEC. This format also makes it easier to understand how far the boundaries are. Let's look at the changes.

A loose-fill opening at or above grade for an underground tank is a Division 2 area that extends up to 18 inches for a horizontal distance of 10 feet from the opening. Where screw-on or quick-connect fittings are used on the fill opening, the horizontal radius is reduced to 5 feet. If the fill pipe is located in a recessed box, or there is a pit or box within the classified area previously defined, they are in a Division 1 area. You will notice that any pits, boxes, or spaces below grade which are within a Division 1 or 2 area are classified as being in a Division 1 location.

The space within a dispenser up to a height of 4 feet is a Division 1 area unless there is a solid partition or solid nozzle boot in the dispenser. If this is the case, the Division 1 area extends up to the barrier and the space above is Division 2. Some pumps with light-emitting-diode readout units come under this category. Since the phrase "nozzle boot" is used here and not defined, an explanation of the term is in order. This is the recessed portion of the dispenser into which the nozzle fits when not in use. "Nozzle holder" or "nozzle storage compartment" are other terms that can be used to describe a nozzle boot. It is not that portion of the flange or cone that is part of the nozzle assembly used with vapor-processing systems, but that part of the dispenser into which the open end of the nozzle is stored when fuel is not being dispensed.

A Division 2 area extends for 18 inches horizontally all around the outside of the dispenser from the upper boundary of the Division 1 area to grade. An 18-inch horizontal dimension extends outward from the nozzle boot, but not around a 90° or greater angle corner; this is also a Division 2 area. Where the dispenser is outdoors, a Division 2 area extends horizontally in all directions outward to 20 feet. The height of this area is 18 inches above grade.

In the previous edition of the NEC there were no dimensions for indoor installations of dispensers. Now there are. Where adequate mechanical ventilation prevents an accumulation of vapor–air mixtures in concentrations above one-fourth of the lower flammable limit, the hazardous (classified) area is the same as for outdoors. With natural ventilation the Division 2 area extends 25 feet beyond the dispenser instead of 20 feet.

The special enclosure for tanks required by Paragraph 7-2.2. of NFPA 30 must be substantially liquid- and vapor-tight. The sides, top, and bottom must be made of reinforced concrete at least 6 inches thick with openings in the top only. The total space within such an enclosure is a Class I, Division 1 location.

Sales, storage, and restrooms are ordinary (unclassified) locations unless any opening into these rooms is within a Division 1 area; then the entire area from floor to ceiling is Classified as Division 1.

All the classifications (Divisions 1 and 2) for vapor-processing equipment are new. Dispensing devices incorporating vapor recovery are treated the same as dispensers as far as extent of classified areas is concerned. Additional areas classified as Division 1 and Division 2 apply to vapor-processing equipment, pits and enclosures containing the equipment, and vacuum-assist blowers used as part of the system. Where the equipment is located aboveground, the Division 2 space extends for 18 inches all around and to the ground, then horizontally outward for a distance of 10 feet to a height of 18 inches above grade.

Although the requirements for overhead dispensing units are not new, there is now more detail. All the space within the dispenser plus 18 inches in all directions of the dispenser enclosure is a Division 1 area. However, the area does not extend beyond walls or ceiling. All electrical equipment integral with the nozzle must be suitable for Class 1, Division 1, Group D locations. The Division 2 area extends for a distance of 2 feet out from the Division 1 area all around the overhead dispenser, then vertically to grade. Assuming that the dispenser housing measures 2 × 2 feet, the Division 1 area extends outward at the pump to a 5 × 5 foot space. The Division 2 area at the ground measures 9 × 9 feet. Note that the Division 2 area extends 3½ feet beyond the nearest edge of the pump on all sides. The Division 2 area also extends 18 inches above the driveway for a distance of 20 feet from all edges of the dispenser. In the example just cited, the distance out is 16½ feet beyond the Division 2 area previously mentioned.

For dispensers with outdoor remote pumps, any pit, box, or space below grade within 10 feet horizontally of any edge of the pump is a Division 1 location. Any space within 3 feet in all directions, and for a height of 18 inches extending out for a distance of 10 feet, is classified as Division 2. Where the pump is installed indoors all space within any pit, regardless of how far away, is a Division 1 location. Division 2 extends 5 feet around the pump in all directions, and to 3 feet above grade extending horizontally for 25 feet in all directions from any edge of the pump.

Where Class 1 liquids are dispensed in a lubrication or service room, any unventilated pit is a Division 1 area; a pit with ventilation is Division 2. Also, Division 2 extends upward for 18 inches above the lubrication pit, extending 3 feet horizontally from its edges. The points of filling and discharging Class I liquids from a dispenser are Division 2 areas that extend in all directions for 3 feet.

Class I liquids have the following characteristics: IA has flash points below 73°F and boiling point below 100°F, IB has flash points below 73°F and boiling point at or above 100°F; IC has flash points at or above 73°F and below 100°F. Although Class I liquids are subdivided according to flash point and boiling point, practically all Class I liquids dispensed at a service station are listed under Group D in Table 500-2.

Where there is no dispensing of Class I liquids in a lubrication or service room, the pit area is classed Division 2, which extends 18 inches above the pit

Section 515-2 109

and for a horizontal distance of 3 feet beyond the sides of the pit. Formerly, all the area in the pit was a Division 1 location and all the remaining space in the room was a Division 2 location which extended 18 inches above the floor.

SECTION 514-8. Rigid nonmetallic raceway is permitted for underground wiring provided that it is buried under not less than 2 feet of earth. To make it clear that nonmetallic conduit cannot emerge from the ground or be exposed in a pit, the last 2 feet of the underground run must be rigid or intermediate metal conduit. Since these raceways must be sealed with seal-off fittings and rigid nonmetallic conduit is not suitable for wiring aboveground in Class 1 locations, it cannot be exposed at either end.

Article 515
Bulk Storage Plants

SECTION 515-2. This Section is completely revised to conform with NFPA 30-1977, *Flammable and Combustible Liquids Code.* Since this committee has responsibility for classification of hazardous areas relative to electrical equipment, the revisions are made to conform with areas defined for bulk plants. Some areas now classified are new in this edition, whereas others have been reclassified. Only the revisions and additions are mentioned.

Where tank cars or tank vehicles are loaded from the bottom and atmospheric venting is used, a Division 1 area extends for 3 feet in all directions from the vent. In the 1978 NEC this was a Division 2 location. The space between 3 feet and 15 feet from the vent which extends in all directions is a Division 2 area. The Division 2 area around the loading connection point is unchanged.

There is no Division 1 area around a closed-dome loading system using vapor recovery, and the Division 2 location only extends outward in all directions from these connection points for a distance of 3 feet.

Formerly, there was no mention of bottom loading with vapor recovery. Now an area within 3 feet in all directions from these connections is listed as a Division 2 location.

Pumps, bleeders, withdrawal fittings, metering, and similar devices, whether installed indoors or outdoors, are in Division 2 locations. The only difference between an indoor and outdoor application is the extent of the Division 2 area. Previously, there was also a classification for inadequately ventilated indoor locations. Since NFPA 30 requires ventilation for indoor installations of pumps and withdrawal fittings, unventilated areas cannot exist.

A Division 1 area exists for all pits and spaces below the floor in buildings used for storage and repair of tank vehicles. Pits and depressions in floors were not previously mentioned. Garages for other than tank vehicles are classed as ordinary locations. Article 511 should be reviewed if other than routine maintenance is performed.

Drainage ditches, separators, and impounding basins are Division 2 locations for a height of 18 inches above these items, and 18 inches above grade, for a horizontal distance of 15 feet from any edge.

Indoor and outdoor drum storage areas are not classified. However, where there is drum and container filling, the volume within 3 feet of the fill and vent openings is Division 1. Between 3 and 5 feet in all directions of the vent and fill openings is Division 2, and a horizontal radius of 10 feet for a height of 18 inches above grade from these openings is also Division 2.

Any pit with mechanical ventilation other than those previously mentioned is an ordinary location unless any part of the pit is within a Division 1 or 2 location because it is within a classified area created by a tank, drainage ditch, pump, etc.

It should be noted that any areas listed in the Table do not extend beyond any unpierced wall, roof, or other solid partition.

SECTION 515-5. Rigid nonmetallic conduit is permitted for underground wiring in Division 1 and 2 areas provided that it is buried under not less than 2 feet of earth. However, rigid metal or threaded steel intermediate conduit must be used for the last 2 feet of the underground run. This change is similar to the one that applies to service stations.

Article 516
Finishing Processes

SECTION 516-2. In Part (b)(5) the extent of the Division 2 area for dip tanks and drain boards is increased from 20 to 25 feet. This change is made to obtain agreement with NFPA 34—Dip Tanks.

Article 517
Health Care Facilities

SECTION 517-1. The third paragraph is added to reaffirm the statement appearing in Section 110-1, which points out that explanatory material is included as Fine Print Notes. Any Standards mentioned in the NEC are not enforceable unless adopted by law or ordinance.

SECTION 517-2. An alternate power source can be a battery system where permitted. Batteries were not recognized previously. This and other changes in definitions are made to obtain agreement between the NEC and NFPA 76A-1977, *Essential Electrical Systems for Health Care Facilities*.

Section 517-13   111

In addition to task illumination and selected receptacles, special power circuits are allowed on the Critical Branch.

The definition of Essential Electrical Systems is reworded with no change in intent.

A definition for "selected receptacles" is added because this term is used to designate equipment permitted on the Critical Branch. The word "minimal" is important to alert designers and users to the fact that unnecessary equipment on the Critical Branch can reduce the reliability of this important system. In other words, reliability should not be downgraded solely for convenience.

SECTION 517-11. Metal faceplates for switches and receptacles are considered adequately grounded where fastened to the device with metal screws that secure the plate to a grounded mounting yoke or grounded outlet box.

Nonmetallic cable is permitted for wiring clinics, medical and dental offices, outpatient facilities, and some nursing homes and residential custodial care facilities provided that the installation of NM cable is allowed by Sections 336-3 and -4. Section 336-3 defines permitted and not-permitted uses for NM cable, including the restriction to structures not exceeding three floors above grade. Section 336-4 states that all other applicable rules in the NEC apply to this wiring method.

Since the 1978 Edition required an insulated equipment-grounding conductor to be included with the circuit conductors supplying receptacles in these occupancies, type NM cable was not often used. In fact, many enforcing authorities required metal raceways because an insulated copper grounding conductor installed with the branch-circuit conductors was required where receptacles or fixed equipment was supplied. This is still the basic requirement of Part (a) of this Section, but new Exception No. 2 relaxes this rule, with a resulting cost reduction for wiring these occupancies.

Part (b) is modified by the addition of the phrase "likely to be used in areas intended for patient care" to make it clear that ordinary appliances used in offices and other nonpatient areas do not have to be grounded unless other Sections of the NEC require grounding. Table and floor lamps, coffee makers, TV sets, typewriters, and calculators are some of the types of appliances that do not have to be grounded under the revised wording.

SECTION 517-13. Receptacles with insulated grounds are permitted in some portions of a hospital. However, they are not acceptable in the patient vicinity and critical care areas. These receptacles cannot be used in the areas mentioned because of the possibility of a difference in potential that could exist between the grounding connection on the isolated ground of this receptacle and the reference grounding point. In critical care areas the maximum difference in potential between any two exposed conductive surfaces cannot exceed 40 millivolts when measured across a 1000-ohm resistance. Receptacles with insulated grounds usually have long ground-return paths, making it difficult to stay within the 40-millivolt limit. For this reason they are not permitted in certain areas of a hospital.

PART D. Nursing Homes and Residential Custodial Care Facilities is rewritten to be in closer agreement with NFPA 76A-1977, *Essential Electrical Systems for Health Care Facilities.* A Fine Print Note is added to reference this Standard.

SECTION 517-40. This Section is rewritten to clearly state which rules apply to nursing homes and residential custodial care facilities. The Exception points out that free-standing buildings housing these occupancies do not have to comply unless medical treatment and examinations are provided or hospital care is included. To be exempt the facility must not accept any tenants who need to be sustained by electrical or mechanical life-support equipment, and must not offer surgical treatment requiring general anesthesia. A facility meeting these restrictions must have a 4-hour emergency supply for selected lighting, nursing stations, medical preparation areas, boiler rooms, communication areas, and alarm systems but does not have to meet other requirements in Sections 517-44 through -47.

SECTION 517-41. Nursing homes and residential custodial care facilities that provide hospital services must meet the requirements for hospitals as spelled out in Part E.

SECTION 517-42. An Essential Electrical System in a hospital can also be used for nursing homes and residential custodial care facilities where contiguous.

SECTION 517-44. The title is changed from "Emergency System" to "Essential Electric System" because the system actually supplies more loads than required by Article 700—Emergency Systems. The Essential Electrical System for nursing homes consists of two branches. The Life Safety Branch generally supplies exit and emergency lighting and can be compared to the emergency systems described in Article 700. The Critical Branch supplies task illumination, special power circuits, and selected receptacles related to patient care.

Usually, two automatic transfer switches are necessary—one for each branch. However, in a small facility one transfer switch is acceptable. A one-line diagram showing this arrangement is shown in Figure 517-44(3).

Part (c) requires adequate capacity in the Essential Electrical System to supply all loads connected to each branch. This does not mean that all loads have to be energized at the same time. For example, the Essential Electrical System must have the capacity to temporarily operate any elevators that may

Section 517-46

have stopped between floors, but full capacity for the heating system does not have to be added to the elevator load, since it is assumed that electric heat will not be supplied during the time the elevators are operating.

The Life Safety Branch is treated the same as emergency lighting in that these circuits cannot occupy the same raceway with any other wiring except in transfer switches, exit and emergency lights supplied from two sources, and junction boxes attached to these fixtures which are supplied from two separate sources.

Since the Critical Branch is not quite as important as the Life Safety Branch, Critical Branch wiring can occupy raceways which contain conductors that are not part of the Branch.

SECTION 517-45. The Life Safety Branch must be energized from the alternate source within 10 seconds after failure of the normal supply. Within this time period it must be able to pickup all the loads specified. Critical Branch loads can be added after the alternate source is stabilized provided that two or more transfer switches are used.

Part (a) contains more detail as to where exit and emergency lighting is required. Also, in patient care hallways, the overhead lighting can be turned off at night where other lighting at a reduced intensity is provided and the switching arrangement is designed only to permit changing from one group of lighting fixtures to another.

The types of alarm and alerting systems that have to be connected to the Life Safety Branch are listed in Part (c).

Any communication systems used for issuing instructions or status reporting must be supplied from the Life Safety Branch. These communication systems can be audible or visual and may include telephone, nurses' call, intercom, paging, security, etc.

Sufficient illumination in dining and recreation areas to adequately indicate exit passageways, and lighting around the generator set are required by Parts (e) and (f). Lighting and power at the generator location are necessary to properly monitor the equipment and perform minor maintenance procedures while the machine is running.

No equipment other than that specified is allowed on the Life Safety Branch. As mentioned earlier, adding unnecessary wiring and equipment to this Branch increases the exposure to failures and reduces the reliability of the system.

SECTION 517-46. Where the Critical Branch is connected to a separate transfer switch as shown in Figures 517-44(1) and (2), a time delay must be employed to allow the Life Safety Branch to be energized first. Task illumination and selected receptacles in patient care areas, sump pumps, domestic water pumps, equipment required to operate for protection of major apparatus, elevator cab lighting, and communication systems are all required to be automatically supplied on failure of the normal source; heating equipment and elevators are allowed to be manually connected to the alternate power source.

In many nursing homes and residential custodial care facilities, individual room air conditioners with strip heaters or heat pumps with supplementary electric heat are used in patient rooms. In other words, central heating and cooling does not exist. In these homes it is not necessary to supply each individual unit from the Critical Branch provided that an area is designated to accommodate all tenants and this area is heated. If the administrator designates the dining room, recreation room, or another area as temporary living quarters, the selected areas must be heated. And the patient room air conditioners do not have to be supplied from the Critical Branch. Some other changes are made to clarify the method used to arrive at an outside design temperature of 20°F. In areas of the United States where the outside design temperature is above 20°F, heating systems are not required to be connected to the Critical Branch.

Under Part (b)(2) it is now mandatory to connect elevators to the Critical Branch; however, a selector switch that only energizes one elevator at a time is acceptable. This arrangement, together with permissive reduction in heating load as mentioned earlier, results in a reduction in the size of the auxiliary power source.

SECTION 517-47. This new Section requires at least two independent sources of power for the Essential Electrical System. Similar language is used in Section 517-65 for hospitals.

According to Part (b), the alternate source must be a generator, but there are Exceptions. The first Exception permits the alternate source to be a utility service or another generator where the normal source is a generator set. And the second Exception recognizes batteries as the alternate source where the nursing home or custodial care facility limits its services by not providing any treatment or care mentioned in the Exception to Section 517-40.

There are no specific requirements for the location of generators, transfer switches, distribution panels, and other components of the Essential Electrical System, but designers, installers, and users are cautioned to give careful consideration to the location of this equipment so that a failure does not occur because of floods, icing, earthquakes, storms, and other natural disasters. Vandalism and sabotage should also be considered when selecting a location for this important electrical equipment. Part (d) also points out that the Essential Electrical System should be designed so that internal wiring and equipment failure will not result in complete loss of power. Coordination studies of overcurrent protective devices at various points throughout the system will show where corrections have to be made so that a completely coordinated system can be obtained. In a properly coordinated distribution system, only the overcurrent device next upstream from the fault will open, while all other circuits not involved remain energized.

Section 517-60   **115**

Part (e) advises that dual sources of normal power should be considered in the design stage because this arrangement provides a more reliable electrical distribution system. Should such a system be contemplated, the two utility company sources have to be designed and arranged so that any fault in the facility will not result in loss of both power supplies.

PART E.   Hospitals has also undergone a complete rewrite, with revisions to remove any conflicts with NFPA 76A-1977 and to more closely track the wording used in that Standard. The reference to NFPA 20 in the Fine Print Note is changed to reflect the latest edition.

SECTION 517-60.   The Essential Electrical System for a hospital is made up of two separate supply sources connected to two subsystems known as the Emergency System and the Equipment System. This terminology is different from that used to describe the systems for nursing homes and residential custodial care facilities but should not cause any difficulty in application as long as the requirements for hospitals are not applied to nursing homes, or vice versa.

The Emergency System is composed of the Life Safety Branch and the Critical Branch. Notice in paragraph (a)(4) that these two Branches must be separate and independent of each other and all other wiring. In other words, each Branch must be in its own separate raceway system; they cannot occupy the same raceway, nor can they occupy any raceway containing any wiring that is not connected to the same Branch. You will remember that Critical Branch circuits for nursing homes and residential custodial care facilities are permitted to occupy the same raceway with other wiring—not so for hospitals.

The enforcing authority has the responsibility for limiting the amount of electrical equipment that may be connected to the Life Safety Branch and Critical Branch of the Emergency System. It is expected that some judgment must be used in the selection of loads permitted on the Emergency System, but remember that the integrity and reliability of the system have first priority.

Part (a)(3) contains a general statement which outlines the purpose and function of the Equipment System, while paragraph (4) allows this system wiring to occupy the same raceway as general-purpose wiring.

Two diagrams [Figures 517-60(1) and 517-60(3)] show different wiring diagrams for small hospitals. In Figure 517-60(1) two transfer switches are shown, one for the Emergency System and the other for the Equipment System. Only one transfer switch is used in Figure 517-60(3), but the title is the same. These diagrams will cause confusion and varying interpretations. First, a "small" hospital is not defined, and then a decision has to be made as to whether a single transfer switch provides sufficient reliability for operation of the equipment connected to it. The decision to go with one transfer switch is further complicated by the statement in paragraph (5) which advises that the number of transfer switches used must be based upon reliability, design, and load considerations. Simply stated, the reliability of the Essential Electrical System is enhanced where more than one transfer switch is specified.

Part (b) says the Essential Electrical System must have adequate capacity to meet the demand for operation of all functions and equipment connected to it. Note that system capacity is based on maximum demand, not necessarily connected load.

SECTION 517-61. The Emergency System consists of two branches: the Life Safety Branch and the Critical Branch. Both Branches must be energized within 10 seconds after loss of the normal source.

These two Branches must be in separate raceways and the wiring must be independent of all other wiring except in transfer switches, in junction boxes attached to exit and emergency lights supplied from two sources, or in these fixtures where supplied from two separate sources. Although not mentioned as an Exception in Part (b), where one transfer switch is permitted for small hospitals, both Branches occupy the same raceway until separated at distribution panels.

Individual branch circuits are required for each isolated power supply serving anesthetizing locations and other areas using isolated power systems. These isolated systems must be supplied from the Critical Branch.

All Life Safety Branch and Critical Branch wiring must be installed in metal raceways. The two Exceptions allow cord-and-plug connections on utilization equipment and wiring without raceways for secondary circuits of transformers supplying communication circuits.

SECTION 517-62. Lighting, alarms, signals, communications, and receptacles that are required to be connected to the Life Safety Branch are listed in (a) through (e). Any equipment providing functions not included in this list cannot be connected to this Branch.

SECTION 517-63. More judgment can be exercised in selecting loads to be supplied from the Critical Branch. Most of the areas or functions listed in Part (a) are not new; none has been removed, but a few are added. Receptacles in psychiatric bed areas cannot be connected to the Critical Branch. Blood, bone, and tissue banks; telephone equipment room and closets; selected acute care beds; selected emergency room treatment areas; and selected postoperative recovery rooms are either new areas or functions, or modifications are made to previous requirements. For instance, all emergency room treatment areas were formerly required to be connected to the Critical Branch. Now only selected areas are required to be so connected.

Additional task illumination receptacles and power circuits needed for effective hospital operation are allowed on the Critical Branch. Justification for the necessity of connecting any additional equipment should be furnished to the authority having jurisdiction by the hospital administrator.

Part (b) permits multiple transfer switches and feeders on the Critical Branch. The possibility of subdividing loads by using additional transfer switches, feeders, distribution panels, etc., should be considered so that any failure during emergency conditions does not result in complete loss of power on the Critical Branch.

SECTION 517-64. This Section has undergone a complete editorial rewrite with revisions. Part (a) contains a list of equipment that must be automatically transferred to the alternate power source after a short time delay. The Fine Print Note advises that staggered starting of the required equipment is acceptable and should be considered so that a grossly oversize generator is not warranted.

An additional time delay to energize equipment mentioned in Part (b) is acceptable. Since the maximum amount of time delay is not specified, it can be assumed that the Equipment System does not have to be supplied until all the load on the Critical Branch is connected to the alternate source.

Environmental heating is required for areas outlined in Part (b)(1). Patient rooms must also be heated from the Equipment System where the outside design temperature is lower than 20°F. However, individual patient rooms do not have to be heated if selected areas are provided for patients and these areas are heated. I am not sure that this part of the Exception has any real application in a hospital because it is difficult to predict how many patients are ambulatory or can be moved at any given time.

Where the hospital is supplied by a dual source of power, either two utility services or two generators, the patient room heating does not have to be connected to the Equipment System

One or more elevators must be connected to the Equipment System so that the hospital can function during a normal power source failure. Other elevators must be connected to the Equipment System at least temporarily so that anyone in an elevator stopped between floors can be removed.

The Equipment System must also supply hyperbaric and hypobaric facilities, automatically operated doors, minimal electrically heated autoclaving equipment and other loads necessary to keep the hospital functional, subject to approval by the authority having jurisdiction. Some examples are kitchen equipment, refrigeration, and laundry machines.

SECTION 517-65. Acceptable sources of power as covered by this Section are generally the same as outlined in Section 517-47. The only additional comment necessary is to point out that batteries are not acceptable as an alternate power source for hospitals, but they are for limited care nursing homes and custodial care facilities.

SECTION 517-80. This Section classifies patient care areas and specifies performance criteria. Formerly, all exposed conductive surfaces except small objects and movable nonelectrical equipment had to be grounded. Now only exposed conductive surfaces likely to become energized have to be grounded with an equipment grounding conductor. Metal doors and frames, window frames, curtain tracks, etc., are some objects no longer requiring grounding where it is unlikely that they will become energized.

General patient care areas are more clearly defined in Part (b)(1). Although invasive procedures such as those resulting from connection of drainage pumps and peripheral intravenous lines are included, these areas are not classified as critical care areas. To properly classify patient care areas, the procedures permitted in general patient care areas and critical patient care areas should be clearly understood, then discussed with the governing body of the health care facility.

The definition of a wet location as applied to patient care areas is revised and added here because these areas are specially treated in Section 517-90(c).

SECTION 517-81. Grounding performance in critical care areas is improved by reducing the allowable voltage drop across a 1000-ohm resistance from 100 millivolts to 40 millivolts. This change should not create any hardship because tests conducted at older hospitals with metal conduit systems and no equipment grounding conductor have shown that potential differences of 20 mV or less are obtainable.

SECTION 517-83. The words "where inpatient hospital care is provided" are added to make it clear that two branch circuits are not required at each patient bed in outpatient facilities and other occupancies listed in Exception No. 2.

SECTION 517-84. A patient equipment grounding point is no longer mandatory at each patient bed in critical care areas. Where equipment grounding conductors in cord-and-plug-connected medical appliances are properly maintained and receptacles are grounded through an insulated grounding conductor, redundant grounding with external grounding conductors plugged into jacks is usually unnecessary.

Any exposed conductive surfaces within 6 feet of the perimeter of, and 7 ½ feet above the floor at the nominal location of the patient bed must be grounded if it is likely that these surfaces can become energized. The Exception lists small fixed metal surfaces, which do not have to be grounded. Large metal surfaces such as door and window frames are added in this edition.

Portable appliances and furniture are added to the list of movable objects that do not have to be grounded through an external conductor connected to a jack at the equipment grounding point.

Section 517-104 119

SECTION 517-90. All 15- and 20-ampere 125-volt single-phase receptacles installed in pediatric locations must be tamperproof either by design or by installation of covers so that children cannot insert metal objects into them.

Since Article 680 now has specific requirements for grounding, bonding, and wiring of therapeutic pools, the Fine Print Note is added.

SECTION 517-101. A gas/vapor-tight metallic sheath is no longer required for type MC cable installed in other-than-hazardous anesthetizing locations according to Part (c)(1).

Hospital-grade receptacles are required for three voltage levels where formerly only 125- and 250-volt ratings were specified.

SECTION 517-103. This Section requiring grounding of electrical equipment in anesthetizing locations is relocated to improve continuity. There is no change in the language.

SECTION 517-104. Yellow is added as the color of the third conductor for three-phase isolated power supplies. With only two colors listed previously, most people assumed that only single-phase systems could be used. This change makes it clear that three-phase surgical tables, X-ray equipment, etc., can be used in anesthetizing locations.

The maximum line isolation monitor leakage current is increased from 2 milliamperes to 5 milliamperes in Part (b)(1). This change is made to agree with NFPA 56A-1978, *Inhalation Anesthetics,* and allows for an increase in circuit capacity or standing leakage.

Lighting fixtures and electrical equipment that shall be or are permitted to be connected to a grounded power supply are covered by Part (c). More detail is provided because the previous requirements were not as clear as they should have been. First, a general lighting circuit supplied from the normal service must supply at least one lighting fixture in each operating room regardless of the type of anesthesia used. The Exception permits storage batteries or battery-powered unit equipment or another source recognized in Section 700-12 where not a part of the Emergency System.

Where fixed lighting fixtures and X-ray equipment are supplied from a grounded system, both must be at least 8 feet above the floor, or the X-ray equipment other than the tube must be located out of the hazardous (classified) space. All surgical lighting fixtures must be supplied from the isolated power supply, and any wiring connected to the isolated system cannot occupy raceways containing any other wiring. Finally, switches controlling grounded circuits either have to be outside the hazardous (classified) area or be wired for remote control with a circuit voltage not exceeding 24 volts.

In other-than-hazardous operating rooms, the rules are generally the same except that X-ray equipment does not have to be 8 feet above the floor.

PART K. X-Ray Installations is rewritten and expanded to cover all the requirements for the installation of medical X-ray equipment. Although the Scope Section of Article 660 previously stated that medical and dental X-ray equipment were not covered, some of the rules did in fact cover this equipment. To eliminate this apparent inconsistency and make Article 517 complete, requirements for medical X-ray equipment are removed from Article 660 and placed in Article 517.

SECTION 517-151. A No. 4 copper conductor was formerly specified for bonding the stationary portion of the X-ray system to the frame of the patient support. Since this wire size may not be adequate to maintain a low-impedance bond for long runs, the size is removed and reference is made to Section 517-81 for critical care patients. This is the Section that limits the potential difference to 40 mV measured across a 1000-ohm resistance.

Article 518
Places of Assembly

SECTION 518-1. Places of Assemby covered by this Article are listed in column format for improved comprehension. There are no additions to or deletions from the list.

Article 520
Theaters and Similar Locations

SECTION 520-21. Dead-front stage switchboards must be constructed according to Part C of Article 384. This refers to standard construction for ordinary occupancies. Any other construction not meeting these provisions must be tested under recognized standards by a qualified testing laboratory.

SECTION 520-45. The requirement that receptacles supplying arc lamps be rated not less than 50 amperes and be supplied by No. 6 conductors is deleted. Now the receptacle has to be rated for 125 percent of the operating current, and the wiring to the receptacle must conform to Article 310. This change recognizes the fact that many presently available arc lamps operate at 20 amperes or less.

SECTION 520-53. Part (f) is updated to recognize solid-state dimmers and other types which do not produce as much heat as resistance dimmers. Conductor insulation has to have a temperature rating of 200°C where resistance-type dimmers are used in a portable switchboard, but for other types of dimmers the temperature rating for the internal wiring is reduced to 125°C.

Article 530
Motion Picture and Television Studios and Similar Locations

SECTION 530-18. The first sentence in Part (d) is new and points out that copper is the only conducting material suitable for cables and cords supplied through plugging boxes.

Article 540
Motion Picture Projectors

There are no substantial changes in this Article.

Article 545
Manufactured Building

SECTION 545-12. A minor change in the wording makes it clear that provisions must be made during manufacture to route the grounding electrode conductor from the service equipment to the point of attachment to the grounding electrode.

Article 547
Agricultural Buildings

SECTION 547-1. Part (c) is added to clarify the status of agricultural buildings not subject to the dust or corrosive environments mentioned in Parts (a) and (b). In my opinion the only time that Part (c) is applicable is where the buildings are not totally enclosed and environmentally conditioned.

Article 550
Mobile Homes and Mobile Home Parks

SECTION 550-2. Definitions "appliance, fixed" and "appliance, stationary" are added here because they still have meaning in this Article. In mobile homes many appliances are secured in place for over-the-road movement. Because this is the case, some appliances could be erroneously classified as fixed or stationary.

SECTION 550-3. The point of entrance of the feeder assembly is no longer restricted to the rear one-third section (opposite end from coupler) of the mobile home. This restriction is removed from Part (i). The practice of installing service pedestals along the rear of the mobile home sites has worked very well where the mobile home feeder is to the rear of the structure. With this restriction removed, mobile home manufacturers have more flexibility in design, but the wiring system in the mobile home park will probably become more complex because of the uncertainty as to where the feeder—in various types of homes—is located.

SECTION 550-23. The requirement that the point of entry of the feeder into the mobile home be not more than 30 feet away is revised to say that the point of attachment of the feeder cannot be more than 30 feet from the mobile home service equipment. This change is necessary because you can no longer predict that the feeder will be located in the last third of the structure. Where the service is in the rear and the feeder entry into the mobile home is toward the front, it is necessary to attach the feeder to some part of the mobile home within 30 feet of the service equipment. From this point the feeder is secured to the structure up to the point where it enters the mobile home.

Article 551
Recreational Vehicles
and Recreational Vehicle Parks

SECTION 551-2. Definitions "portable, fixed" and "stationary appliances" are added because many cord-and-plug-connected appliances are secured in place to prevent movement during over-the-road travel. Although these terms do not have any significance in this Article, they should be of some help in meeting requirements for disconnecting means and overcurrent protection of appliances as covered by other Sections of the NEC.

SECTION 551-3. Part (c)(5) is new and requires the chassis grounding terminal of the battery to be bonded to the vehicle frame with a No. 8 copper conductor. This assures an adequate-size grounding path back to the battery terminal should a fault develop on the ungrounded dc conductor.

A cover or equivalent protection must be installed over low-voltage fuses to prevent accidental short-circuits.

SECTION 551-4. In determining voltage converter ratings, low-voltage appliances controlled by momentary contact switches are not considered connected loads if the equipment is used to prepare the vehicle for occupancy or travel. Electrically operated jacks, exterior curtains, and retractable entrance steps are some examples.

The converter enclosure must be bonded to the vehicle frame with a No. 8 copper conductor. A ground lug is usually provided on the converter for this purpose. This size wire provides sufficient ampacity for a ground return path should the ungrounded dc conductor contact the frame.

SECTION 551-5. The supply conductors from the engine generator can now be directly connected to a panelboard mounted on the outside of the generator compartment wall.

SECTION 551-8. There is no restriction on the number of circuits permitted on a 30-ampere distribution panel and feeder assembly. In the 1978 Edition, only one 20-ampere air-conditioner circuit and two 15- or 20-ampere circuits were allowed. This change in Part (c) correlates with a revision in Section 551-11(c), which now requires a main overcurrent protective device not exceeding the rating of the Feeder Assembly where more than two circuits are supplied.

SECTION 551-11. Working clearances in front of panelboards are now defined in Part (b). Because of the limited space in recreational vehicles, the minimum clearances listed in Table 110-16(a) are not practical. Furthermore, it is usually not necessary to work on this equipment while it is energized.

The Exception permits an additional reduction of working space. Where the panel board is exposed to an inside aisle, one of the clearances can be reduced to 22 inches. Since most aisles are 22 or more inches wide, this does not create any problem. Under this condition the space in front of the panelboard can be reduced to 24 inches wide and 22 inches deep.

The revision in Part (c) requires main overcurrent protection where more than two circuits are in use. This revision was previously mentioned in discussing the changes in Section 551-8.

SECTION 551-13. The proper terminology for a male-recessed motorbase receptacle is "flanged surface inlet."

The length of the cord is no longer restricted to 26 ½ feet. Only a minimum length of 20 feet, or a length equal to the distance from the point of entrance to the front of the recreational vehicle, is specified.

SECTION 551-14. A wood backing plate at least ½ inch thick surrounding the outlet box opening and extending 1 ½ inches beyond the box on all sides is an acceptable mounting means. Both of the dimensions are actual and not nominal lumber dimensions. This Exception to Part (e) is necessary because structural wall and ceiling members for supporting boxes do not exist in some types of recreational vehicle construction.

Article 555
Marinas and Boatyards

SECTION 555-1. The Scope is amended to include moorage of floating dwelling units (houseboats).

SECTION 555-20. Although the Scope Section states that Part B applies to the moorage of floating dwelling units, this Section and Section 555-21 can be interpreted to mean that interior wiring is also covered. But that is not the intent.

SECTION 555-22. Overhead service conductors must be installed so that minimum clearances are maintained at the lowest water level.

SECTION 555-23. The wiring between the houseboat and shore must be designed so that there is no strain on the conductors because of changing tides and wave action.

SECTION 555-24. The floating dwelling unit and the electrical distribution system must be connected to an earth ground on shore.

SIX
Special Equipment

Article 600
Electric Signs
and Outline Lighting

SECTION 600-2. Switches, flashers, etc., controlling transformers must be rated for inductive loads or have twice the ampere rating of the transformer primary current. The Exception allows ac general-use snap switches to be used up to their ampere rating.

SECTION 600-4. Formerly, all electric signs had to be listed. Now listing can be waived by special permission. Unless the sign is built on the site or is an unusual "one-of-a-kind" type, written consent from the enforcing authority stating that listed signs are not required in his jurisdiction will be hard to come by.

SECTION 600-5. Exception No. 2 recognizes bonding of isolated noncurrent-carrying metal parts of outline lighting (individual letters are a good example) with No. 14 wire if protected from physical damage. Section 250-94 contains similar language and the change here brings the two into agreement.

SECTION 600-6. Part (b) requires an outlet on a 20-ampere branch circuit for each commercial occupancy with ground-floor footage. Although this circuit has been required for a number of years, there was no mention of load calculations for this circuit. Now Part (c) specifies a load of 1200 volt-amperes. This valve should be adequate for most outdoor signs installed for small commercial occupancies.

SECTION 600-11. Ground-fault circuit-interrupters are required for all readily accessible portable signs located outdoors after January 1, 1982. The GFCIs must be installed on or in the mobile or portable sign to protect anyone touching the frame from serious electric shock. Since these trailer-mounted signs are moved from one commercial location to another, receive little maintenance, are exposed to the elements, and are subject to vandalism, it is necessary to provide more protection from electric shock than is available through the cord-and-plug grounding connection. Although a sign-mounted GFCI does not protect against the hazards of a defective cord, to require that the sign be connected to a GFCI-protected receptacle is even more difficult to enforce. This revision might not be the complete answer to protection of people from electric shock, but I believe enforcement will be enhanced by requiring GFCI protection integral with the sign.

SECTION 600-31. Sign conductors operating at voltages above 600 can now be installed in rigid nonmetallic conduit. This material is added to the list of acceptable wiring methods.

Article 604
Manufactured Wiring Systems

This is a new Article which deals with a plug-in type wiring system that has been around for a few years. Cable in various lengths is assembled with connectors on the ends for plugging into terminal boxes and lighting fixtures. This wiring method is designed for branch-circuit wiring above accessible ceiling spaces.

SECTION 604-1. Wiring assemblies are manufactured with termination fittings on one or both ends of the cable for plugging into multioutlet boxes, lighting fixtures, and receptacle assemblies. It can also be used for signaling and communication circuits.

SECTION 604-2. Special-configuration receptacles and connectors are secured to the cable during manufacture and cannot be removed and reinstalled in the field.

SECTION 604-3. This wiring system can only be used in accessible and dry locations. If the cable is a type that is permitted in plenums and air-handling spaces used for environmental air, and the product is listed for this application, it may be installed in these areas.

The Exception allows a switch leg to be concealed in a hollow wall. Cables prepared for use with switches usually have a connector on one end with wires sticking out of the armor at the other end for termination in a switchbox.

SECTION 604-4. If the preassembled cable is type AC (armored cable), it cannot be used where prohibited by Section 333-6. For type MC cable the rules in Section 334-4 apply. Some manufactured wiring systems use flexible metal conduit as the wiring method.

SECTION 604-5. All other applicable rules must be satisfied during installation of this wiring system. Particular attention should be given to the support requirements for armored cable and metal-clad cable. Type AC cable must be supported every 4 ½ feet and within 12 inches of every outlet box or fitting. Six feet is the maximum distance between supports for type MC cable.

SECTION 604-6. Each cable, whether type AC or MC, must contain a No. 12 bare copper equipment grounding conductor. Where the cable assembly is flexible metal conduit, the equipment grounding conductor must be insulated. The circuit conductors must be No. 12 copper. No other size is recognized.

According to Part (b), all receptacles and connectors must be of the polarized and locking type. Receptacles and connectors (male and female) must have different configurations so that a connector of one voltage rating cannot be plugged into a receptacle of different voltage. For example, a 120-volt connector has to be of a design that will prevent it from being plugged into a 277-volt receptacle. Communications and signal circuit terminations must also be of a different design to prevent mistakes. All connectors and receptacles must be specifically polarized and designed for system, voltage, and function to prevent mismatching subassemblies of different systems.

Lighting fixtures, power poles, communication outlets, etc., must be listed to use with the wiring assembly. Lighting fixtures are provided with one or two integral connectors mounted on the top of each fixture for connection to a distribution receptacle box and for interconnection between fixtures. The distribution unit is a junction box containing receptacles and the branch-circuit wiring from the panelboard. This is the part of the system that supplies power to the plug-in wiring system.

SECTION 604-7. Any male and female fittings that are not used must be capped.

Article 610
Cranes and Hoists

SECTION 610-2. Part (c) is added to point out the special requirements applicable to cranes and hoists operating above electrolytic cells. It is essential that some conductive parts be insulated from ground to protect workers in the cell area from hazardous electrical conditions.

SECTION 610-31. A minor revision makes it clear that the disconnecting means for runway conductors must be readily accessible from the floor or ground.

Article 620
Elevators, Dumbwaiters, Escalators, and Moving Walks

SECTION 620-12. With the increased use of semiconductors and electronic and logic circuits, the minimum wire size is changed from No. 20 to No. 24 for operating control and signal circuits.

SECTION 620-21. The wording is changed to make it clear that rigid nonmetallic conduit is not acceptable as a wiring method for this equipment.

SECTION 620-53. Phase protection for hydraulic elevators is now covered by Part (b). Note that there is no requirement to prevent starting the motor as is required for polyphase ac drive elevators. Since there is no immediate danger of hydraulic elevators moving in the wrong direction as is the case with phase reversal on polyphase ac motor drives, it is necessary only that the hydraulic pump motor be protected from overheating because of phase reversal or phase failure.

SECTION 620-91. The first paragraph was inadvertently omitted from the 1978 Edition of the Code. This language is the same as that which appeared in the 1975 Edition and requires energy absorption on the load side of the elevator power disconnecting means to limit car overspeed to the valves specified.

Article 630
Electric Welders

There are no changes in this Article.

Article 640
Sound-Recording and Similar Equipment

There are no substantial changes in this Article.

Article 645
Data Processing Systems

SECTION 645-1. The words "in a data processing room" are added at the end of the sentence to exclude remote display units such as telephone terminals used in department store billing and inventory control, and cash registers in supermarkets.

Additional wiring methods are listed in Part (c)(2) and "metal" is added to "rigid conduit" to prohibit nonmetallic conduit. Metal wireways are added to provide more circuit capacity in a given space. Large computer installations require many branch circuits which can best be installed in wireways. Multiple runs of wireways under raised floors are a practical way to handle circuit additions and relocations whenever data processing units are added or relocated.

Type AC (armored cable) is now an acceptable wiring method under a raised floor.

SECTION 645-3. This Section is rewritten for clarification of intent. The disconnecting means must disconnect all electronic equipment in the computer room. It does not have to disconnect lighting, office machines, and other appliances. Another disconnect is required for the air conditioning system. The location is now specified as the principal exit doors instead of the old phrase "at designated exit doors." There is no longer any requirement for a disconnecting means when data processing equipment is located in general building areas. Since the Scope is revised to indicate that this Article applies only to data processing rooms, any reference to equipment in general office areas or other locations is not in agreement with the Scope and must be deleted.

Article 650
Organs

There are no revisions in this Article.

Article 660
X-Ray Equipment

There are no substantial changes in this Article.

Article 665
Induction and Dielectric Heating Equipment

SECTION 665-1. The Scope is changed to indicate that induction heating of pipelines and vessels at line frequency is not covered. This is new Part E in Article 427, which was previously discussed.

SECTION 665-64. Part (c), which required grounding of the center point of a coil in a tank or chamber where a vacuum or controlled atmosphere is used, is deleted because leak-detection devices connected to these melting furnaces do not function properly.

Article 668
Electrolytic Cells

SECTION 668-30. Auxiliary electrical devices such as motors, transducers, alarms, etc., mounted on electrolytic cells can be wired with multiconductor hard-usage cord as well as other wiring materials.

Article 669
Electroplating

The wiring methods employed in electroplating systems are not in compliance with general rules found in the NEC. This is because some of the requirements do not fit the unique installation methods needed by the plating industry. This new Article is the result of a Technical Subcommittee report and should enhance worker safety.

SECTION 669-1. This Article covers electrical installations used for electroplating, anodizing, electropolishing, and electrostripping.

SECTION 669-2. Where applicable, general rules in Chapters 1 through 4 govern except as modified by this Article.

SECTION 669-3. All electrical equipment used in the electroplating process must be suitable for the specific purpose, function, use, environment, application, etc.

SECTION 669-5. Branch-circuit conductors must be sized for 125 percent of connected load. The ampacity of copper bus is 1000 amperes per square inch and for aluminum 700 amperes per square inch.

SECTION 669-6. Electroplating conductors operating at 50 volts dc or less shall either be insulated or if bare must be supported on insulators. Above 50 volts dc insulated conductors have to be supported on insulated supports and protected from physical damage. Bare conductors must be supported on insulated supports and protected from accidental contact by enclosures, screens, or by placing at an elevation above the working surface.

SECTION 669-7. Warning signs have to be posted in the vicinity of the bare conductors to alert everyone to the hazard of energized circuits.

SECTION 669-8. The disconnecting means for the dc wiring is permitted to be a switch, removable links, and removable conductors.

SECTION 669-9. Overcurrent protection for the dc conductors shall be fuses or circuit breakers, or current-sensing equipment that will open the disconnecting means. Other types of current-limiting systems which limit the current, reduce the output voltage, deenergize the dc conductors, or in some other way prevent overloading are also acceptable.

Article 670
Metalworking Machine Tools

There are no changes in this Article.

Article 675
Electrically Driven or Controlled Irrigation Machines

SECTION 675-9. A revision is made to make it clear that the equipment-grounding conductor does not have to be any larger than specified in Table 250-95. Before this new third sentence, the old rule was interpreted to require an equipment-grounding conductor equal in size to the supply conductors.

SECTION 675-22. The method used for selection of branch-circuit conductors, controllers, and disconnecting means is changed so that motor nameplate information is applied in the calculations rather than the full-load currents listed in Table 430-150.

SECTION 675-23. Instead of being on the center pivot irrigation machine, the disconnecting means can now be located within 50 feet provided that it is visible from the controllers on the machine. In addition, the disconnecting means must be capable of being locked in the open position regardless of whether it is on the machine or 50 feet away.

Article 680
Swimming Pools, Fountains, and Similar Installations

SECTION 680-1. The Scope is expanded to include hot tubs and spas. The Fine Print Note is added to clarify the meaning of pools and fountains.

SECTION 680-4. Some definitions are modified to include fountains. For example, a dry-niche lighting fixture may be installed in a pool or fountain.

An Exception which stated that therapeutic pools in health care facilities are not covered is removed because these pools are now covered by Part F.

A definition for a spa or hot tub is added because rules are added for installation of this equipment.

The dimensions for storable pools are increased from 15 feet to 18 feet for the maximum dimension and from 3 feet to 3 ½ feet for the maximum depth. Most of the storable pools are more than 3 feet deep. This change recognizes current manufacturing practice.

SECTION 680-6. The Exception is revised to permit a receptacle supplied by 240 volts to be installed for a pump without GFCI protection for the receptacle. Formerly, there was no voltage level assigned to receptacles located within 15 feet of the inside walls of a pool. Because of the wording, a 240-volt circuit for a cord-and-plug-connected pump had to have the single locking and grounding receptacle protected by a ground-fault circuit-interrupter. The revised wording eliminates this requirement.

A receptacle was required not closer that 10 feet or more than 15 feet from the inside walls of a pool "at an existing dwelling." These words are removed because they only caused confusion.

The Exception for Part (b) is revised by removing the requirement for GFCI protection of existing lighting fixtures located less than 5 feet horizontally from the inside walls of a pool. Although it is no longer necessary to install a GFCI for these lighting fixtures, they have to be grounded in accordance with Article 250—Grounding.

SECTION 680-8. The Table and Diagram are modified to indicate that vertical clearances for overhead utility wires extend for 10 feet beyond the inside walls of the pool or to the outer edge of diving structures, observation stands, towers, or platforms, whichever is greater.

The second Exception is new and allows utility-owned, -operated, and -maintained communication wires and CATV cables to pass over pools and related structures where there is not less than 10 feet of vertical clearance.

SECTION 680-10. Any underground wiring foreign to the pool must be at least 5 feet away from the inside walls of a pool.

The Exception allows a lesser distance where space limitations do not permit 5 feet clearance. However, the underground wiring must be installed in corrosion-resistant rigid metal conduit, intermediate metal conduit, or nonmetallic conduit. Burial depths must be in accordance with Table 300-5 for the wiring method used. If metal raceway is used, bonding to the pool reinforcing steel is necessary to comply with Section 680-22(a)(5).

SECTION 680-11. This is old Section 680-47, which formerly appeared under Part D—Fountains. Since the rule is applicable to all swimming pools, fountains, and similar installations, it is moved to Part A—General. Now there is no doubt that the requirement applies to all installations covered by Article 680.

SECTION 680-20. Part (a)(4) is revised to say that inherent protection against overheating is necessary for lighting fixtures that depend on submersion for safe operation. A low-water cutoff was previously mentioned.

SECTION 680-25. Revisions to the wording and relocation of some Parts are made for better understanding. Panelboards are added in Part (a) because grounding rules for this item are covered in Part (c).

Part (d) is revised to recognize connection of electrical equipment other than underwater lighting fixtures by any of the wiring methods of Chapter 3. Recirculating pumps and pool water heaters are examples of equipment that can be supplied underground with type UF cable where located more than 5 feet from the inside walls of the pool and buried according to Section 300-5.

SECTION 680-27. This is a new set of rules recommended by the Technical Subcommittee—Deck Heating of Swimming Pools. The requirements cover electric heating installed within 20 feet of the inside walls of a pool.

Unit heaters must be totally enclosed or guarded types, rigidly supported to the structure, and located more than 5 feet from the inside walls of the pool.

Radiant heaters must be permanently installed and cannot be less than 12 feet above the deck or closer than 5 feet horizontally to the inside edges of the pool.

Heating cables in the deck are not allowed.

SECTION 680-40. Spas and hot tubs installed outdoors must comply with the general rules of Part A and Permanently Installed Pools—Part B, unless modified by the Exceptions.

Metal bands or hoops used to secure wooden staves do not have to be bonded nor is additional bonding required where metal-to-metal connections are made between various parts.

Package spas and hot tubs that are listed by an independent testing laboratory do not have to be disassembled to check bonding and grounding, because bonding and grounding must be done at the factory in order to obtain listing or labeling. These package assemblies can be cord-and-plug-connected to a GFCI protected receptacle.

SECTION 680-41. For indoor spas and hot tubs, listed package units are allowed to be cord-and-plug-connected. All others assembled on the site must be permanently connected to a wiring method recognized in Chapter 3.

No receptacles are permitted within 5 feet of a spa or hot tub. Any receptacles more than 5 feet and less than 20 feet from the spa or tub must be protected by GFCIs. Spas and hot tubs supplied through flexible cord must be plugged into a GFCI protected receptacle.

Lighting fixtures installed above or within 5 feet horizontally of the inside edges of a spa or hot tub must be connected to a circuit protected by GFCI. Wall-mounted switches are not permitted within 5 feet of the inside walls of the spa or hot tub unless bonded to the metal structure.

Part (d) requires bonding of all metal parts of the spa or hot tub. All metal parts of equipment associated with the spa or hot tub must be bonded, as well as metal conduit, piping, metal surfaces, and any electrical devices, controls, and equipment that are within 5 feet of the inside walls. Recognized bonding methods include threaded pipe connections; metal-to-metal bolted, welded, or riveted joints; interconnection between metal parts with a No. 8 solid copper wire.

Grounding of all electrical equipment associated with the circulating system, including pumps and blowers and all electrical equipment within 5 feet of the inside walls of the spa or hot tub, is required by Part (f).

Methods of grounding must be in accordance with Article 250 for the wiring system used. Part (2) of (g) requires the equipment-grounding conductor of a flexible cord to be connected to a fixed part of the spa or hot tub assembly.

SECTION 680-52. A rewrite of Part (b) is made to clarify the requirements. All underwater junction boxes and other enclosures are required to be watertight. They must be of corrosion-resistant material and firmly attached to the supports or directly secured to the fountain surface. However, junction boxes may be supported by corrosion-resistant metal conduit.

SECTION 680-60. This is the beginning of a new set of rules dealing with therapeutic pools and tubs in health care facilities. Portable therapeutic appliances which are often used in patient rooms for treatment of small areas of the body include an equipment-grounding conductor in the power supply cord or are double-insulated. Hence, they are not covered by these rules but are required to comply with Article 422—Appliances.

SECTION 680-61. Permanently installed therapeutic pools must comply with the rules of Parts A and B except that lighting fixtures above and around the pool do not have to be protected by GFCIs if the fixtures are totally enclosed.

SECTION 680-62. Large therapeutic tanks which are not moved must have GFCI protection for all associated electrical equipment.

The bonding requirements for all electrical parts, metal piping, etc., are covered by Part (b) and are similar to those for hot tubs and spas, which were previously discussed in detail.

Part (c) covers methods of bonding and is similar to Part (e) of Section 680-41. Connection by suitable metal clamps is a method not mentioned for spas and hot tubs.

There is no difference between the grounding requirements in Part (d) for therapeutic tubs and Part (g) of Section 680-45 for spas and hot tubs.

All receptacles within 5 feet of the inside walls of a therapeutic tank have to be protected by a ground-fault circuit-interrupter to comply with Part (f).

SECTION 680-63. Only totally enclosed lighting fixtures are permitted in the therapeutic tub area. Although the tub area is not defined, I assume that it extends horizontally outward for a distance of 5 feet from the inside walls of the tub or tank.

Article 685
Integrated Electrical Systems

SECTION 685-1. This is a new Article dealing with industrial processes which must be shut down in an orderly manner to minimize hazards to personnel or equipment. The rules also apply to any other operations that could be hazardous should a power failure occur.

To qualify under these special rules, the electrical equipment must be maintained by qualified persons and the authority having jurisdiction must be satisfied that adequate safeguards are enforced.

SECTION 685-2. For ready reference, a list of Sections and Exceptions dealing with orderly shutdown is included. These include overcurrent protection, grounding, and motor overloading.

SECTION 685-10. Overcurrent devices (fuses and circuit breakers) can be mounted more than 6 ½ feet above the floor or working surface. Since a disconnecting means is required on the supply side of cartridge fuses, disconnect switches are also covered by this rule.

SECTION 685-12. Two-wire dc systems do not have to be grounded, but are required to be equipped with a ground detector to comply with Exception No. 1 of Section 250-3(a).

SECTION 685-14. Ungrounded separately derived control circuits are permitted by Exception No. 3 of Section 250-5(b). Here again a ground detector is required. Actually, ungrounded systems do not provide any more reliability than a grounded system if accidental grounds are not removed when they occur. A ground detector lets you know that a ground has developed so that appropriate corrective action can be initiated.

SEVEN
Special Conditions

Article 700
Emergency Systems

SECTION 700-1. The Scope is rewritten to more clearly define the purposes and functions of an emergency system. Note that emergency systems are installed primarily to provide illumination and power to loads considered essential for safety to human life. There is no reference to protection of property as previously appeared in this Scope because new Article 701—Legally Required Standby Systems is added for this purpose.

Two Fine Print Notes are added to alert code users to the fact that additional rules apply to this system where used in health care facilities.

The former second paragraph of this Section contained explanatory material only but appeared in regular print. It is now a Fine Print Note with revisions. Items such as "essential refrigeration, operation of mechanical breathing apparatus, illumination and power for hospital operating rooms," etc., are removed because they are either not essential for the safety of human life or are covered by rules in other parts of the NEC or other Standards. Items added include elevators and public safety communication systems.

SECTION 700-4. The emergency system must be tested under load according to Part (e). The type of equipment needed to make this test and its location are no longer specified. This revision allows the designer more freedom in selection of the testing method and location of the test equipment.

SECTION 700-5. Part (b) is new and permits use of the alternate power source for nonessential loads during times when an emergency does not exist. Examples of loads that may be supplied are air conditioning, heating, ventilation, lighting, and pumps. When the alternate source is operating in this mode, automatic transfer and switching must be provided to drop this load and pickup the emergency system within 10 seconds and the legally required standby system (if it exists) within 60 seconds.

Where the alternate emergency power source is used for selective load pickup and shedding, a portable or temporary power source must be available whenever the emergency generator is out of service for extended periods of time. If more than one generator set is used for emergency and legally required standby power and the others have adequate capacity to supply the essential loads, a temporary or portable unit is not necessary.

Major maintenance or repair includes engine overhaul, piston and piston ring replacement, crankshaft and bearings removal, and similar operations requiring engine or generator disassembly. It does not include oil and filter changes, spark plug replacement, radiator flushing, and similar routine maintenance procedures.

SECTION 700-6. Transfer switching arrangements have to be listed or labeled for use on emergency systems or be approved by the authority having jurisdiction. The transfer equipment design must prevent inadvertent interconnection of the normal and emergency sources. This language does not prohibit interconnection of the two systems where proper paralleling arrangements are part of the alternate power source.

SECTION 700-7. Signals, audible and/or visual, should be provided to indicate that the emergency source is not operational; the battery or generator set is carrying load; the battery charger is not functioning, and the prime-mover starting equipment is not available for starting the prime mover. This last item is new. On many generator sets the starting equipment is controlled through a manual "start–off–automatic" selector switch. Incidents have been reported where the engine-generator did not start in an emergency because the selector switch was not returned to the automatic position after routine testing.

Although signal devices for the various functions mentioned are not mandatory, they should be considered as essential to assure availability of the emergency power source. High-engine-temperature and low-oil-pressure alarms should also be provided, even though not listed in Section 700-7.

Section 700-12   143

SECTION 700-9. This is old Section 700-17 with revisions. It should be obvious by now that Article 700 has undergone a major rewrite and relocation of various Sections to improve continuity and understanding.

All wiring for the emergency system must be kept independent of any other wiring. The Exceptions relax this requirement to the extent that mixing of normal and emergency wiring cannot be avoided.

The fourth Exception is new and allows more than one emergency circuit in the same raceway, while Exception No. 5 permits normal supply conductors and output conductors from a unit equipment to occupy the same raceway. Since a unit equipment battery pack is allowed to serve remote lamps, the dc wiring to these lamps can be routed through the same junction box with the ac supply conductors provided that insulation on the dc conductors is suitable for 600 volts.

SECTION 700-12. This was formerly Section 700-6, with which everyone is familiar. The major change in the first paragraph is to limit the time to 10 seconds between failure of the normal supply and reenergizing the emergency loads from the auxiliary source.

The third paragraph is added to alert designers and users to the fact that a reliable emergency system is also dependent on its location. Where these natural and man-made hazards cannot be avoided, other precautions should be taken, such as connecting sump pumps to the emergency system where the alternate source is located in a basement, or heating the area from the alternate source where icing and freezing weather are encountered.

The revision in Part (b)(2) recognizes so-called "day tanks" in the building, with larger tanks located underground or outdoors.

Day tanks for diesel engines are used for fuel transfer from an underground storage tank where the engine-driven fuel pump does not have the capacity to draw the fuel from the main supply tank. Usually, an electrically driven auxiliary pump is provided to transfer fuel from the underground tank to the day tank, which is located close to the prime mover. Day tanks vary in capacity from about 5 to 60 gallons.

Incidents have been reported where the electric auxiliary fuel pump was not connected to the emergency system, causing the engine to stop when the fuel supply in the day tank was consumed. Make sure that all electrically driven equipment associated with the prime mover is connected to the emergency system.

Where a natural gas engine is the prime mover, an on-site fuel supply is no longer necessary if there is low probability of simultaneous loss of the natural gas supply and power from the electric utility. Before making the decision to eliminate the on-site fuel supply, consideration must be given to the effects of hurricanes, earthquakes, flooding, and other natural disasters, explosions, and curtailment of natural gas supplies.

Municipal water supplies are no longer recognized as the sole cooling medium for internal combustion engines because the water supply may not be available when needed, and untreated water running through an engine causes rapid corrosion of internal parts, thus reducing reliability of the emergency system.

Since the emergency load must be picked up in 10 seconds, Part (b)(5) is added to allow a short-time-rated power unit to supply emergency power until the prime mover can accelerate to operating speed. Large prime movers such as turbines may take longer than 10 seconds to reach operating conditions.

Uninterruptible power supplies are recognized as an auxiliary power source by Part (c) provided that the batteries comply with Part (a) and the charging equipment meets the requirements of Part (b).

A separate service is still recognized as an alternate power source in Part (d). However, permission to use this arrangement must be obtained from the authority having jurisdiction.

Permission to tap ahead of the service disconnecting means must be granted by the authority having jurisdiction . This method of obtaining an alternate source is probably the least reliable of any recognized in this Section. If this arrangement is used, the connection for the alternate source cannot be made in the main service disconnecting means. Two suitable connection points are at the service head for overhead services and at an outdoor junction box for underground services.

SECTION 700-26. This Section is added to clarify the intent of Section 230-95, which requires ground-fault protection of equipment on 480Y/277-volt services having service disconnecting means rated at 1000 amperes or more. Where engine-generator sets at this voltage exceed this current (a 900-kW generator, for example), ground-fault protection is not mandatory. The rule is written to let the electrical engineer evaluate all the advantages and disadvantages of ground-fault protection for the generator before making a decision. Although many variables have to be weighed, one that deserves primary consideration is how much value should be placed on preventing equipment damage where human safety is involved.

Article 701
Legally Required
Standby Systems

This is a new Article, which includes most of the requirements that previously appeared in Article 750—Standby Power Generation Systems. Much of the work on Articles 700, 701, and 702 was done by a Technical Subcommittee of Code Making Panel 22. The original proposal by the committee was to combine Articles 700 and 750, but the National Electrical Code Correlating Committee directed that the Articles remain separate. The Code Making Panel then created three Articles, with one following the other. Hopefully, a better understanding of the requirements as they apply to each particular system will result.

SECTION 701-1. Legally required standby systems must be permanently installed in their entirety. Additional rules for these systems where installed in health care facilities are found in Article 517 and NFPA 76A-1977.

SECTION 701-2. Legally required standby systems must automatically supply designated loads upon failure of the normal supply.

The Fine Print Note gives examples of the types of loads that may be required to be supplied. The list is not all-inclusive, since local laws and ordinances vary. The phrase "could create hazards or hamper rescue or fire fighting operations" is the basic reason for this system.

Because many of the requirements in this Article are similar to those in Article 700, only the major differences are mentioned.

SECTION 701-5. The inspection authority is not required to witness periodic testing after the initial test is observed.

SECTION 701-6. The standby source only has to have sufficient capacity to supply all equipment intended to be operated at one time. This is in contrast to the emergency system, which must have adequate capacity to operate all connected equipment.

SECTION 701-9. Legally required standby system wiring is permitted to occupy the same raceway with other wiring.

SECTION 701-10. The maximum time for transfer from the normal supply to the alternate source is 60 seconds.

As with Emergency Systems, municipal water supplies are not acceptable as a cooling medium for generator sets. A utility gas supply does not have to be supplemented with an on-site fuel supply where there is low probability of loss of normal electric service and the gas supply at the same time.

Article 702
Optional Standby Systems

SECTION 702-1. Only engine-generator sets permanently installed are covered by this Article. However, precautions must be taken with portable units that are temporarily connected to a premise wiring system. The main hazard associated with this equipment is caused when there is a back-feed into the utility system.

SECTION 702-2. Since an optional standby system is not required by law, the owner is free to choose the kilowatt rating of the unit and the loads to be served. The Fine Print Note mentions some loads that are typically connected to optional standby systems and explains that the purpose is to reduce discomfort and prevent serious interruption of the process or damage to the product.

SECTION 702-5. The alternate source must have sufficient capacity to supply all loads intended to be energized at one time. By alternating electrical loads, the rating of the generator set does not have to be any greater than the largest load served.

SECTION 702-6. Automatic or manual transfer equipment is acceptable. However, it must be designed and connected to prevent inadvertent interconnection of the alternate and normal sources.

SECTION 702-7. Audible and visual signals are not required but may be considered necessary for some applications.

SECTION 702-8. There are no special requirements for optional standby circuits. Wiring methods recognized in Chapter 3 are acceptable unless not permitted by other Sections of the Code.

Article 710
Over 600 Volts,
Nominal—General

SECTION 710-2. Busways are now a recognized wiring method for systems operating above 600 volts. Article 427 is added to the list because skin-effect current tracing for pipelines and vessels may be connected to medium-voltage (over 600 volts) systems.

SECTION 710-3. Medium-voltage cables can now be installed in rigid nonmetallic conduit. Previously, this wiring method was not recognized for above ground use.

Unshielded cables are now allowed to be directly buried if the cable assembly meets the requirements of Section 310-7 or if the construction of the cable satisfies one of the Exceptions to Part (b).

The first Exception recognizes type MC cable and the second Exception permits lead sheath cable. To be acceptable, the metallic sheath of the cable must have the capacity to conduct safely any fault current likely to be imposed on it and have sufficiently low impedance to operate the overcurrent protective devices in the circuit.

Part (c) is added to include copper or aluminum busbats as a wiring method.

SECTION 710-9. Metal-enclosed switchgear and industrial control assemblies are added to this Section, which prohibits foreign piping and ducts above and around medium-voltage equipment. The former rule applied only to service equipment.

SECTION 710-72. The value of the neutral and ground currents for electrode-type boilers is increased from 5 amperes to 7 ½ percent of boiler full-load current for 10 seconds, or an instantaneous value of 25 percent of boiler full-load current.

Under normal operating conditions the neutral circuit conductor carries very little current, but unequal phase loading does occur from steam bubbles forming in the resistance path between electrodes, uneven wearing of electrodes, and nonuniform buildup of boiler scale on electrodes. Because of these conditions, neutral currents sometimes exceed 5 amperes.

The revision has little effect on electrode boilers with full-load currents of about 65 amperes or less because 7 ½ percent of full-load current in not more than 5 amperes. However, the instantaneous current can be as great as 16 amperes. For larger electrode boilers the allowable neutral current increases proportionally, but branch-circuit wire size must also increase, resulting in decreased resistance in conductors. Since conductor resistance decreases as neutral current increases, the voltage drop in the neutral conductor which appears on the boiler structure is not a shock hazard.

Article 720
Circuits and Equipment Operating at Less than 50 Volts

There are no substantial changes in this Article.

Article 725
Class 1, Class 2, and Class 3 Remote-Control, Signaling, and Power-Limited Circuits

SECTION 725-2. Part (e) is added to indicate that some motor control circuits should be judged under Part F of Article 430. Any control wiring tapped from the load side of a motor branch circuit must meet the requirements of Section 430-72(a).

SECTION 725-3. The Fine Print Note is added to explain the reasoning behind the voltage and current limitations for Class 2 and Class 3 circuits. Simply stated, Class 2 circuits are those that do not present a fire or shock hazard, while Class 3 circuits are not considered a fire hazard but do present a shock hazard.

SECTION 725-12. The second Exception is new. Generally, Class 1 conductors connected to the secondary of a transformer are required to have overcurrent protection. However, where the secondary is only two-wire single-voltage type, secondary overcurrent protection may not be required where the combination of primary overcurrent protection and transformer secondary-to-primary voltage ratio does not allow any more current than that permitted by Part (b).

To show how this works, let's assume a 1000 volt-ampere 240-volt single-phase transformer with a two-wire secondary that is rated 30 volts. With 5-ampere fuses as the overcurrent protection on the primary, the maximum secondary current is 40 amperes. This example shows that control circuit overcurrent protection is required. By performing the calculations as indicated in the text and considering that Nos. 18, 16, and 14 are properly protected from overcurrent where protection does not exceed 20 amperes, we apply the transformation ratio of (30 ÷ 240), then multiply this figure by 20 to arrive at a maximum primary overcurrent protective device of 2.5 amperes. If this size overcurrent device is used in the primary, the control conductor sizes mentioned earlier do not need secondary overcurrent protection, but the maximum output of the transformer is reduced to 600 volt-amperes.

Control circuit conductor overcurrent protection can be as great as 300 percent of the ampacity of No. 14 and larger conductors. For No. 14 wire, this value is 60 amperes, and for No. 12 copper it is 75 amperes. Here is a practical application of the third Exception: No. 14 copper control wiring run to a remote on–off switch for a 60-ampere lighting contactor without individual overcurrent protection for the control circuit.

SECTION 725-18. Rigid nonmetallic conduit is added to the list of wiring methods that offer protection from physical damage.

SECTION 725-35. The rewrite is for better understanding of the requirements. Overcurrent protection for Class 2 and 3 circuits cannot be interchangeable with devices of higher ampere ratings. These devices may be an integral part of the transformer or other power supply.

SECTION 725-40. The Exception to Part (b)(3) recognizes a metal sheath construction without an outer nonmetallic jacket for type PLTC cable. The construction details in subpart (3) did not anticipate a metal sheath. Therefore, a nonmetallic jacket over the conductors could be enclosed in a metal sheath, but another nonmetallic sheath over the metal was necessary to comply with the rule. Now the Exception eliminates the final outer nonmetallic covering.

Article 760
Fire Protective Signaling Systems

SECTION 760-1. All NFPA Standards are updated to show the latest edition at the time of adoption of the 1981 NEC.

SECTION 760-16. Types KF-2 and KFF-2 are added to Part (b). These are recognized fixture wires rated 600 volts with maximum operating temperature of 200°C.

SECTION 760-22. Electrical supervision of fire protective signaling circuits is not required for household fire alarm equipment unless the signal lines extend beyond the building.

SECTION 760-25. The revised wording for overcurrent protection is the same as used in Section 725-35 for Class 2 and 3 circuits.

SECTION 760-28. This Section and Section 760-29 are rewritten with some interchange of rules from one Section to the other. The main reasons for these changes are to clarify intent and permit nonpower limited wiring methods for power-limited systems. The title is changed to also include materials for power-limited alarm systems. As noted earlier, power-limited circuits are those which have maximum voltages and currents as shown in Table 760-21(a). There are more restrictions on nonpower-limited wiring; therefore, this superior wiring is certainly acceptable for power-limited fire protection signaling circuits. The number of conductors permitted in raceways is the same as for general wiring.

Where nonpower-limited wiring methods are used, conductors do not have to meet the requirements for "building" wire as described in Chapter 3 provided that fixture wires or multiconductor cables mentioned in Sections 760-16 or -17 are installed.

Regardless of the number of wires installed in a raceway, derating of conductors is not necessary for nonpower-limited circuits. Note that Section

760-18(a) does require conductor derating for nonpower-limited wiring where installed with Class 1 circuits.

Subpart (1) to Part (b) is old Section 760-29, which dealt with physical protection of conductors. Where an installation cannot be made as indicated, conductors must be protected in a raceway. The raceway material is not specified, but (2) recognizes metal and rigid nonmetallic conduits for this application. Nonmetallic conduit is not allowed in hoistways, but flexible metal conduit and type AC cable are permitted under restricted conditions outlined in the Exception to Section 620-21.

For vertical runs of conductors from floor to floor, conductors must be encased in noncombustible tubing or be listed as having a fire-resistant covering. Where cables are located in a fireproof shaft, firestops are necessary at each floor.

SECTION 760-29. Remaining portions of old Section 760-28 that are not included in new Section 760-28 are moved to this Section. There is no change in any of the regulations.

SECTION 760-30. The revision here recognizes wiring methods that exceed those required by Parts (a) through (f). In other words, nonpower-limited wiring is acceptable. However, there may be a problem with using some of the flexible stranded fixture wires (type TFF, PFF, etc.) mentioned in Section 760-16(b), because Part (a) and the Exception require solid, seven-strand or bunch-tinned stranded conductors. Fine-stranded fixture wire is not recommended for this application because only one or two strands connected at a terminal are adequate to carry the low value of "supervisory" current but will probably melt from higher current under an alarm condition. An open circuit results and there will be no alarm activation except for "open-trouble" indication.

Exception No. 3 is added to recognize a metallic jacket without an overall nonmetallic jacket. Without this Exception a metal jacket over the thermoplastic jacket has to be covered with another nonmetallic jacket to conform to the requirements of Part (c).

Coaxial cables are added to the types of wires that are acceptable for power-limited circuits. In computerized fire protection signaling systems, low-loss transmission is essential. Before this revision, shielded twisted-pair cables were used which had considerably higher line losses.

EIGHT

Communication Systems

Article 800
Communication Circuits

SECTION 800-2. The reference to grounding communication conductors to a cold water pipe that appeared in Subpart (c)(1)c is removed because revised Section 800-31 more fully explains acceptable grounding methods.

The Fine Print Note is revised to agree with the definition for "effectively grounded" as it appears in the National Electrical Safety Code ANSI C2-1977.

SECTION 800-31. Suitable grounding electrodes for the protector ground are covered in revised Part (b)(5). To make the grounding conductors as short as possible, the connection should be at the nearest accessible component of the grounding electrode system. Where a building or structure has an electric service containing a grounded circuit conductor, the connection points are the metal service conduit, the service-equipment enclosure, or the grounding electrode conductor at the connection to the grounded conductor.

For ungrounded systems without a grounding electrode system described in Section 250-81, suitable attachment points for protector grounding are itemized in Subpart (c). If none of these grounding electrodes exist, the following electrodes are acceptable: metal underground water pipe, metal frame of a building, concrete-encased electrode, or a ground ring. Finally, the grounding electrodes mentioned in Subpart (c) can be used where no others are available.

Notice that accpetable grounding procedures are listed in descending order from most effective to least effective. Short ground leads between the protector and service-grounding system limit the voltage that can develop on the communication wiring and equipment due to lightning and accidental contact between communication equipment and power wiring.

Effective August 1, 1981, an accessible means external to the service equipment enclosures must be provided at all dwellings for grounding communication and CATV cables entering buildings and for all antenna discharge units. This item was mentioned when Section 250-71 was discussed.

A sentence is added to Part (b)(6) which requires clamps, connectors, etc., suitable for direct burial or concrete encasement where used to connect the grounding conductor to a pipe or other electrode.

Article 810
Radio and Television Equipment

SECTION 810-21. Metal masts, supports, and antenna discharge units must be grounded in accordance with Part (f). This language is the same as used in Part (b)(5) of Section 800-31 and needs no further discussion.

Grounding of an antenna installed on a penthouse or similar elevated location can be accomplished by connection to a grounded metal water pipe or rigid or intermediate metal conduit. It is not necessary to run the grounding electrode conductor all the way down to the service equipment.

Article 820
Community Antenna Television and Radio Distribution Systems

SECTION 820-7. The outer conductive shield of coaxial cable must be grounded where exposed to lightning, lightning arrester conductors, or power conductors operating as a potential of over 300 volts to ground. Part (a) recognizes direct grounding of the shield, and Part (b) permits grounding of the shield through a protective device. The protective device is intended to prevent overheating and failure of the shield due to neutral fault currents which may appear on the shield because of a loose or open neutral at the power service, or because of improper bonding of the grounding electrode system and related wiring.

Section 820-22

SECTION 820-14. The text is revised to conform with the wording used in Section 300-21, which deals with the spread of fire and products of combustion.

SECTION 820-22. Part (f) lists suitable grounding electrodes in descending order. If the first and preferred grounding electrode system is not available, go to the next, etc. The outer conductor of a coaxial cable has to be grounded to one of these grounding electrodes. The language used in this revision follows closely the wording in Section 800-31(b)(5) and Section 810-21(f).

NINE

Tables and Examples

NOTES TO TABLES. Dimensions of conductors and the allowable number of conductors in a raceway are based on the nominal size of conductors and conduit or tubing. Actual dimensions of wire may vary from the values given in the Tables. Compact stranded conductors and multiconductor cables are not included. Where these conductors are used, the actual dimensions should be used for determining conduit or tubing fill.

The Fine Print Note is added to alert designers, contractors, and electricians to the fact that using the maximum fill permitted by the Tables may cause difficulty in pulling the conductors into the raceway. This is especially true for 1/2- and 3/4-inch conduits using the maximum number of 14s, 12s, or 10s. Although the figures for these conductors as shown in Tables 3A, 3B, and 3C are correct (not more than 40 percent fill), difficulty is encountered because conductors are not properly aligned, 90° bends are near each end of the raceway, the length of conduit between pull boxes is excessive, etc. Consideration must be given to the total number of bends and the length of conduit between pulling points before deciding to use the maximum fill allowed by the Tables. Selecting a larger-size raceway or reducing the number of conductors can save a lot of time and effort. Proper training or cabling of conductors to prevent crossovers while pulling is an important factor in reducing friction. Wire-pulling compound is also essential.

Dimensions for No. 8 stranded conductors are added to Tables 6, 7, and 8. A few years ago, Section 310-3 was revised to say "where installed in raceways, conductors of size No. 8 and larger shall be stranded." The addition of the physical characteristics of No. 8 stranded to these Tables provides the information needed for calculating raceway fill where No. 8 and other sizes of conductors are installed in the same conduit.

Examples. In Example No. 2 the calculations for the floodlight load are revised by addition of 25 percent because this load is considered continuous. According to Part (c) of Section 210-22 and the definition for a continuous load in Article 100, a continuous load cannot exceed 80 percent of the branch-circuit rating. Multiplying the actual load by 125 percent also results in a minimum branch-circuit rating. One 15-ampere 115-volt branch circuit is needed for the floodlight.

Example No. 3 is revised to include 80 duplex receptacles. The calculations show proper application of demand factors as permitted by Table 220-13. Also, an outside sign circuit is added to comply with Section 600-6 (b). This is the Part that requires a 20-ampere 120-volt circuit at an accessible location outside the building for all commercial occupancies with ground floor space. The load to be added for this circuit is specified as a minimum of 1200 volt-amperes according to Section 600-6(c).

Although the minimum number of lighting circuits is listed for various combinations of ampere ratings and two-wire and multiwire circuits, you should refer to new Part (d) of Section 220-3, which requires only a sufficient number of circuits and branch-circuit overcurrent devices to serve the actual connected load. The service, feeders, and branch-circuit panelboards have to be sized for the computed load, but circuit wiring and branch-circuit overcurrent devices have only to be provided to serve the installed lighting.

The feeder neutral load calculations for the main feeder supplying 40 dwelling units as explained in Example 4(a) are revised to show an additional reduction in demand of 25 percent for the lighting and appliance load which exceeds 120,000 watts. In the 1978 Edition a 35 percent demand was applied to 213,000 watts even though Table 220-11 allows a demand factor of 25 percent to be applied to that portion of the calculated load exceeding 120,000 watts.

Calculations for motor branch circuits and a feeder supplying the motors are covered by Example No. 8. The title is changed to agree with the terminology used to define motor-branch-circuit overcurrent protection (short-circuit and ground-fault) and running overcurrent protection (motor overload).

Sections 430-6 and -7 are added to the list of references because these Sections provide detailed instructions on how to determine motor horsepower rating and full-load current. Calculations for feeder ampacity and overcurrent protection are included in the Example; therefore, Section 430-62 is added.

Tables and Examples **161**

The body of the text is revised by substitution of "short-circuit and ground-fault protection" in place of "branch-circuit overcurrent protection," and "motor overload protection" replaces "motor-running overcurrent protection." These are editorial revisions to correspond with the language used in Articles 430 and 440. As mentioned earlier, "overload" as used in these Articles means current due to motor overloading and failure to start, while the motor-branch-circuit protective device only protects connected equipment from ground-faults and short-circuits.

The motor overload protection for a 30-hp motor is 50 amperes based on 125 percent of full-load current. Since this value can be increased if the motor will not start or carry the load, a reference to Section 430-34 is included.

In selecting the branch-circuit short-circuit and ground-fault protection for a 25-hp motor, an assumption is made that the motor will not start and accelerate its load; therefore, the ampere rating of a nontime-delay fuse is increased to 110 or 125 amperes. These ratings are still less than 400 percent of full-load current as permitted by Exception No. 2a of Section 430-52.